Springer Tracts in Advanced Robotics

Volume 58

Editors: Bruno Siciliano · Oussama Khatib · Frans Groen

Torsten Kröger

On-Line Trajectory Generation in Robotic Systems

Basic Concepts for Instantaneous Reactions to Unforeseen (Sensor) Events

Springer

Author

Torsten Kröger
Technische Universität Carolo-Wilhelmina zu Braunschweig
Institut für Robotik und Prozessinformatik
Mühlenpfordtstraße 23
D-38106 Braunschweig
Germany
E-mail: t.kroeger@tu-bs.de

ISBN 978-3-642-26229-6 ISBN 978-3-642-05175-3 (eBook)

DOI 10.1007/978-3-642-05175-3

Springer Tracts in Advanced Robotics ISSN 1610-7438

Softcover reprint of the hardcover 1st edition 2010

Typeset & Cover Design: Scientific Publishing Services Pvt. Ltd., Chennai, India.

Printed in acid-free paper

5 4 3 2 1 0

springer.com

STAR (Springer Tracts in Advanced Robotics) has been promoted under the auspices of EURON (European Robotics Research Network)

There is always an easy solution to every
human problem — neat, plausible, and wrong.

Henry Louis Mencken (1880–1956)

Foreword

By the dawn of the new millennium, robotics has undergone a major transformation in scope and dimensions. This expansion has been brought about by the maturity of the field and the advances in its related technologies. From a largely dominant industrial focus, robotics has been rapidly expanding into the challenges of the human world. The new generation of robots is expected to safely and dependably co-habitat with humans in homes, workplaces, and communities, providing support in services, entertainment, education, healthcare, manufacturing, and assistance.

Beyond its impact on physical robots, the body of knowledge robotics has produced is revealing a much wider range of applications reaching across diverse research areas and scientific disciplines, such as: biomechanics, haptics, neurosciences, virtual simulation, animation, surgery, and sensor networks among others. In return, the challenges of the new emerging areas are proving an abundant source of stimulation and insights for the field of robotics. It is indeed at the intersection of disciplines that the most striking advances happen.

The goal of the series of *Springer Tracts in Advanced Robotics (STAR)* is to bring, in a timely fashion, the latest advances and developments in robotics on the basis of their significance and quality. It is our hope that the wider dissemination of research developments will stimulate more exchanges and collaborations among the research community and contribute to further advancement of this rapidly growing field.

The monograph written by Torsten Kröger is the outcome of six years of work by the author during his doctoral thesis. The book focuses on sensor integration in robotic manipulation control systems, and in particular on the instantaneous planning of motion trajectories in reaction to unforeseen sensor events, such as failures or more simply a change of reference frame or control space. The supporting theoretical tool is hybrid switched-system control enabling to switch from sensor-guided motion, e.g., under force/torque or visual servo control, to sensor-guarded motion and vice versa. The resulting online trajectory generation algorithm serves as an intermediate layer between low-level motion control and high-level sensor-based motion planning. Numerous examples illustrate the proposed concepts in practice.

This volume is a very fine addition to our STAR series!

Naples, Italy
September 2009

Bruno Siciliano
STAR Editor

Preface

Research activities in the field of engineering cannot be performed by individuals alone. One always has to intercommunicate, to discuss, to exchange knowledge and experiences — to work together. This monograph is the result of six years of research on robot motion control at the Institut für Robotik und Prozessinformatik at the Technische Universität Braunschweig.

It contains nothing but the description of an algorithm for on-line trajectory generation in robotics. The original idea of this algorithm was born during my diploma thesis in 2001. At first, I thought I could develop such an algorithm as a small subpart of my thesis, but after two weeks working on a concept, I decided to simplify the problem and developed an algorithm whose functionality was greatly reduced. During my time as a PhD student, I supervised a number of master students, who worked on a more advanced version of the algorithm, but the algorithms we developed at that time were erroneous and incomplete, as I know now. It took me some years to discover what the actual problem of this trivial and seemingly simple idea was — and the basic idea was extremely simple, not even worth publishing. Now, years later, the original idea has been realized, and the way from the idea to its realization shall be conclusively, compactly, and comprehensively presented in this book. During this project, I have interacted with many people. Each has had some influence on the final version of this document, and I am absolutely grateful for all these contributions.

First of all, I would like to thank my supervisor, Professor Friedrich Wahl. He provided excellent technical equipment, an ideal environment, and the best imaginable atmosphere for a group of young, highly motivated researchers. All the discussions, all the encouragements, and the great latitude during all stages of this work gave me the possibility of working very efficiently and — in particular — of choosing my own area in the field of robotics research. I sincerely appreciate this.

Professor Herman Bruyninckx from the Department of Mechanical Engineering at the Katholieke Universiteit Leuven not only reviewed this work as the second examiner, he has also been an important discussion partner over the years. After our first meeting in 2003, we met sporadically, and we always had great dialogs that were always very fruitful for me.

I should like to give a special word of thanks to the *whole* staff of the Institut für Robotik und Prozessinformatik. The friendly and cooperative environment is preeminent. In particular, I owe a debt to Daniel Kubus. He proofread the manuscript, and he has always been an excellent and competent discussion parter who supported me in manifold ways.

Furthermore, I would like to express my gratitude to my former diploma supervisor and later colleague, Bernd Finkemeyer. He belongs to the most reliable and respectable persons I have met in my whole life. The core of Chap. 7 of this book is based on his ideas, approaches, and experiments. Although he left the university, we still discuss and meet regularly, which very often leads to valuable new ideas.

During the development of the on-line trajectory generation algorithms, a number of mathematical problems occurred, and I am very grateful to Professor Sándor Fekete, Professor Harald Löwe, and Professor Rainer Löwen, all of whom gave me immediate support in a very efficient way.

Michael Marschollek gave me a short introduction to neurophysiology and very good references to this field, such that I could write a subsection about the neurophysiological system of human beings in order to compare the reflexes of humans and robots.

All diploma and master students who worked under my supervision have earned great tribute. Especially, Michaela Hanisch, Christian Hurnaus, and Adam Tomiczek strongly supported me with their efforts, discussions, and ideas. All three worked hard on the first ideas and implementations of the concept of on-line trajectory generation.

Many thanks to all my friends in Braunschweig and in other places — we really had a great time during the past eleven years of study. I will never forget this lovely period of life. Finally, I thank my family for their outstanding aid during all my years of study.

Braunschweig,
January 2009

Torsten Kröger

Abstract

This monograph focuses on sensor integration in robotics, in particular in robotic manipulation control systems. We consider a mechanical system with multiple degrees of freedom equipped with one or more sensors delivering digital and/or analog sensor signals. There is no question that sensor integration and sensor-based control belong to the dominating domains for the future advancement of robotic systems. Although there has been much research on this objective, there is still one important question that has not been answered yet: If we consider a robot in an arbitrary state of motion, how can we calculate a trajectory, if we want the robot to react *instantaneously* to *unforeseen* sensor events?

The core part derives a class of algorithms that generate motion trajectories for robotics systems on-line, that is, within one control cycle (typically one millisecond or less). Such an algorithm is executed in parallel to low-level motion controllers, and systems using it are able to react *instantaneously* to unforeseen (sensor) events. In order to answer the above question, the algorithm enables switchings from sensor-guided robot motion control (e.g., force/torque or visual servo control) to trajectory-following motion control and vice versa, which is an important feature for the practical realization of sensor-based robot motion control systems. The proposed on-line trajectory generation algorithm acts as an open-loop pose controller at an intermediate control layer that functions as one element of the important bridge between low-level robot motion control and higher-level (sensor-based) motion planning. Furthermore, it enables robotic systems to perform a kind of *robotic reflex*.

As the first derivation step, the algorithm is developed for systems with one actuator only, and subsequently it becomes extended to be applicable in robotic systems with multiple degrees of freedom. The generated trajectories are time-optimal *and* synchronized for all degrees of freedom, such that all degrees of freedom reach their target at the same time instant.

Using the proposed on-line trajectory generation algorithm as one control submodule in a hybrid switched-control system simplifies the execution of sensor-guided and sensor-guarded motions. As the algorithm is able to take over control at any time instant and in any state of motion, safe and continuous motions can be guaranteed — even if sensors fail. An additional benefit is that unforeseen (sensor-dependent) switchings of reference frames and/or control spaces become possible.

The proposed concept is of a very basic nature and, thus, addresses various fields of robotics, in which sensor integration plays a fundamental role, for example, in service robotics, manipulation control systems, mobile robotics and manipulation, and robotic surgery. Samples and use-cases accompany the book in order to provide a comprehensible insight into this interesting and relevant field of robotics.

Contents

Chapter 1
Introduction

Sensor integration belongs — without any doubt — to one of the key technologies for the future advancement of robotic systems. This introductory chapter describes the subject of on-line trajectory generation (OTG) and its relation to (multi-)sensor integration and sensor-based control in the field of robot technology. Furthermore, it contains a brief outline of this monograph.

1.1 Robot Motion Control

The field of OTG is not dedicated to one particular kind or group of robotic systems. As will be figured out in the book, OTG is of very fundamental character and applicable to all kinds of robotic systems, in which sensor integration and sensor-based control plays a fundamental role. Hence, a robot is considered a very abstract kind of mechanical positioning system with multiple degrees of freedom (DOF).

We take the definition of the term "robot" specified by the *International Organisation for Standardisation* in ISO 8373 [117] as the basis: "An automatically controlled, reprogrammable, multipurpose manipulator programmable in three or more axes, which may be either fixed in place or mobile for use in industrial automation applications." This definition is also accepted by the *International Federation of Robotics (IFR)* [116] and by the *European Robotics Research Network (EURON)* [75]. Although this work originates from the field of industrial manipulation control, the presented concept is not limited to the field of industrial automation applications. OTG is also relevant in many other fields of robotics, such as service robotics, manipulation control systems, mobile robotics and manipulation, parallel kinematic machines, humanoid robotics, and robotic surgery. Besides, also the development of motion generation algorithms for computer numerical controlled (CNC) machine tools is addressed in this work.

Motion control of robotic systems belongs to the most classical areas of research in the field of robotics. In the following two subsections, we briefly

T. Kröger: On-Line Trajectory Generation in Robotic Systems, STAR 58, pp. 1–9.
springerlink.com

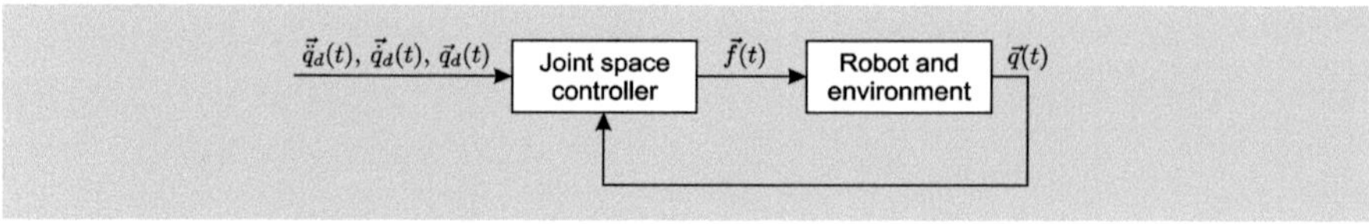

Fig. 1.1 Basic outline of an example scheme for joint space control.

distinguish between trajectory-following control and sensor-guided robot control. Based on this distinction, we introduce a definition of *sensor-guarded* motion control.

1.1.1 Path Planning and Trajectory-Following Operations

One of the simplest methods of robot motion control is trajectory-following control, which is common in industrial manipulation control systems. As shown in Fig. 1.1, command variables containing joint positions $\vec{q}_d(t)$, velocities $\dot{\vec{q}}_d(t)$, and accelerations $\ddot{\vec{q}}_d(t)$ are sent to a joint controller, which is responsible for minimizing the error between the desired robot position $\vec{q}_d(t)$ and the actual $\vec{q}(t)$ in order to achieve a good trajectory-following behavior. Depending on the joint type, rotatory or prismatic, the output of the joint controller $\vec{f}(t)$ contains forces and/or torques for the joints of the mechanical system. The command variables $\vec{q}_d(t)$, $\dot{\vec{q}}_d(t)$, and $\ddot{\vec{q}}_d(t)$ are functions of time and describe the *trajectory* of the robot. In contrast to this, a *path* is only a geometric representation of the desired robot motion without any consideration of time. Hence, we generally have to distinguish between *trajectory generation* and *path planning*. In the field of path planning, we take geometric descriptions of the robot and its static or dynamic environment for granted. The aim is to find a collision-free path from an initial pose to a target pose; this search is commonly done in the configuration space. Fig. 1.2 shows a trivial example, in which the path from the initial position $\vec{p}_{start}$ to the target position $\vec{p}_{trgt}$ is represented by three vertexes, which are connected by splines. The aim of trajectory generation however, is to calculate command variables in order to reach the target position and orientation along the specified path under consideration of some criterion, for example, minimum-time or minimum-energy.

1.1.2 Sensor-Guided Robot Motion Control

In contrast to trajectory-following operations, which commonly do not allow a controlled reaction to sensor signals, sensor-guided robot motions only work on the basis of sensor signals, that is, the sensor is part of the control loop. Fig. 1.3 shows a control scheme on a very abstract level. We take the

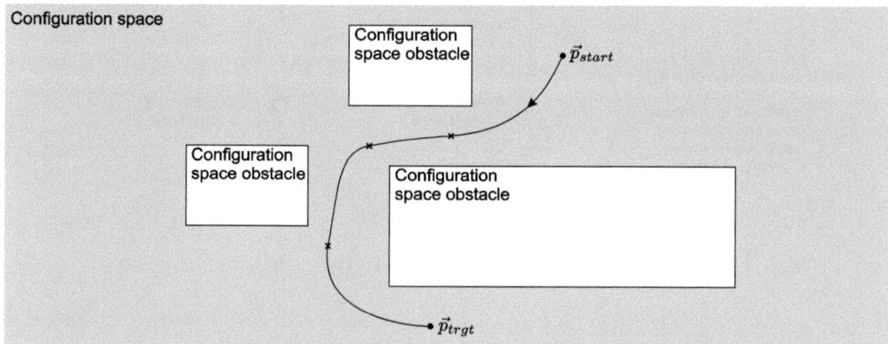

Fig. 1.2 Path planning: two-dimensional configuration space example of a collision-free path.

simple concept of Fig. 1.1 as the basis and assume that the robot is equipped with some sensor system (e.g., a force/torque sensor or a computer vision system) that delivers a signal $\vec{s}(t)$. Here, we consider a *sensor* as a generic device delivering a signal depending on the overall system state, that is, the robotic system with its entire environment. The robot motion of one single control cycle depends on the sensor signal(s) of the current control cycle. The respective controller for this signal acts in task space and uses its command variable $\vec{s}_d(t)$ as well as the position feedback $\vec{p}(t)$ to generate new set-points $\vec{p}_d(t)$, $\vec{\dot{p}}_d(t)$, and $\vec{\ddot{p}}_d(t)$. These values are transformed back into joint space and act as command variables for the joint controller. It is obvious that this system can react to arbitrary situations, states, and events as long as they are detected by the sensor system. The reaction behavior only depends on the transfer function of the controller for the sensor signal $\vec{s}(t)$.

1.1.3 Verbal Problem Formulation and Motivation for This Book

Up to here we have defined the two basic kinds of robot motions: trajectory-following control (without sensors) and sensor-guided motion control. For

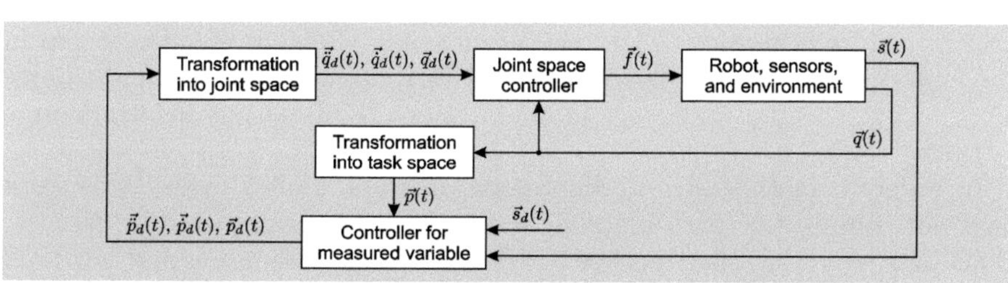

Fig. 1.3 Very abstract example of a scheme for sensor-guided robot motion control.

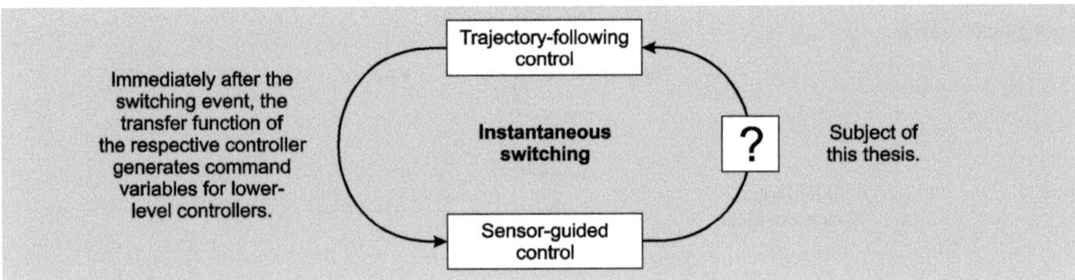

Fig. 1.4 Very simple scheme to illustrate the subject of this book.

the realization of practically relevant systems, both kinds of control are required — of course. It is even necessary to combine both and use them at the same time; then some system DOFs are guided based on some sensor signal, and others are controlled with regard to a specified trajectory. A simple example of a system with three independent translational DOFs would be a motion command, for which force control in z-direction, distance control in y-direction, and trajectory-following control in x-direction is applied. The literature is full of approaches, descriptions, and research on sensor-based control. Force/torque control has had a dedicated community since the beginning of the 1980s as well as the field of visual servo control has since the beginning of the 1990s. The same is of course true for the field of trajectory generation for robotic systems; here the roots can be found in the 1960s. On the other hand, we have to wonder, why these proven concepts and technologies can hardly be found in commercially available robotic systems.

Major problems appear when all these fields are merged together: We must be able to switch abruptly at unforeseen time instants from one kind of robot control to another. If we arbitrarily switch one or more DOFs from trajectory-following control to sensor-guided control, this problem is usually a solved. By using the transfer function of a desired controller, command variables for lower-level control can be generated at any time instant. But how can we switch *one, some, or all* DOFs of a robotic system from sensor-guided control to trajectory-following control? If we consider a robot in an arbitrary state of motion how can we calculate a trajectory if we want the robot to *react instantaneously to unforeseen (sensor) events*?

This is the central question of this work. Its answer allows us to change trajectory parameters for one or more DOFs at any time instant and in any arbitrary state of motion. Thus, we can not only close the significant gap in the loop depicted in Fig. 1.4, but we can furthermore influence and change a currently executed trajectory in one or more DOFs at any arbitrary time instant.

In order to make a strong differentiation from similar fields: This work does not contain any path planning methodologies. As depicted in Fig. 1.5, OTG acts as an interface between high-level motion planning and low-level joint control.

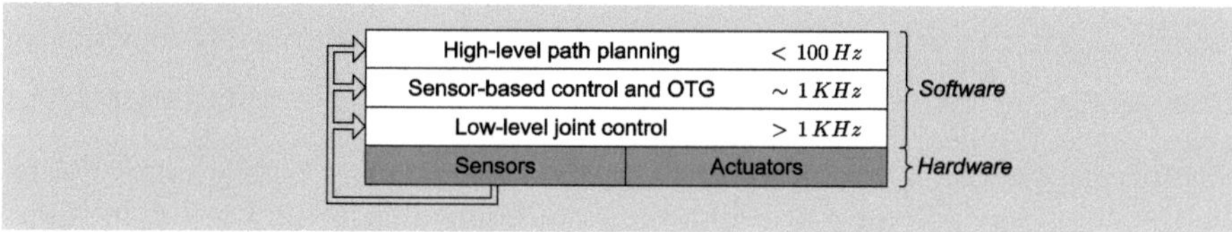

Fig. 1.5 Abstract three-layer model for robot motion control in accordance with [36].

1.1.4 Definition: Sensor-Guarded Robot Motion Control

We consider a robot equipped with one or more sensors delivering digital and/or analog sensor signals. The time instant, in which an analog sensor variable exceeds or under-runs a certain value, a digital sensor output switches, or a Boolean expression consisting of a number of conditions containing any available sensor signal becomes true, is considered a sensor event. A simple example of a sensor event is the transition from free space to contact (or vice versa). As soon as the measured contact force exceeds a certain threshold above the noise level, contact can be detected. Another example is the unforeseen detection of an object (e.g., a human being) by a computer vision system. All these are events that the robot has to instantaneously react to. Thus, we extend our classification to

1. trajectory-following control (Section 1.1.1)
2. sensor-guided motion control (Section 1.1.2)
3. sensor-guarded motion control

and define the latter one as follows:

Definition 1.1 (Sensor-Guarded Motion). *Sensor-guarded motions are considered as trajectory-following motions and/or sensor-guided motions, whose motion parameters for one, some, or all DOFs may change abruptly depending on* sensor events. *These parameter changes may include set-point switchings for closed-loop controllers (e.g., new force/torque control set-points) as well as new desired target positions and/or new constraints for trajectory-following control.*

In the moment of contact detection, for example, the control system may immediately and abruptly switch the respective controllers for all DOFs in contact from trajectory-following control to force/torque control. A second obvious example of the field of visual servo control is the instantaneous reaction of a robotic system to any predictable or unpredictable event: The manipulator of [142] plays the parlor game Jenga [107] and has to interrupt any motion as soon as the game tower topples.

The fundamental achievement of sensor-guarded motions is that robotic systems become enabled to react instantaneously to unforeseen (sensor) events — similar to reflexes of living beings. Due to this absolutely new development, a new philosophy for robot control architecture may become possible. Since this development has rarely been considered in the robotics research literature on control systems technology, the next chapter briefly investigates how the neurophysiological system of human beings performs a reflex motion.

1.2 Excursion: The Neurophysiological System of Human Beings

Sensor-guarded motions are comparable with many everyday human life scenarios: If a little child accidentally touches a stove with its hand, it knee-jerkily reacts by pulling its hand away from the hot surface. Another higher-level human scenario could be the fight between two swordsmen: Depending on the motion of their opponent, the fighters react *immediately* by adapting their own body motions and moves.

This section makes a short excursion from robotics to the human neurophysiological system and explains how a human reflex is performed. Except for the three blocks on the left, Fig. 1.6 was created in accordance with [262]. The author is clearly aware to the fact that the figure is wrong from a strictly neurophysiological or anatomical point of view. The only purpose of this

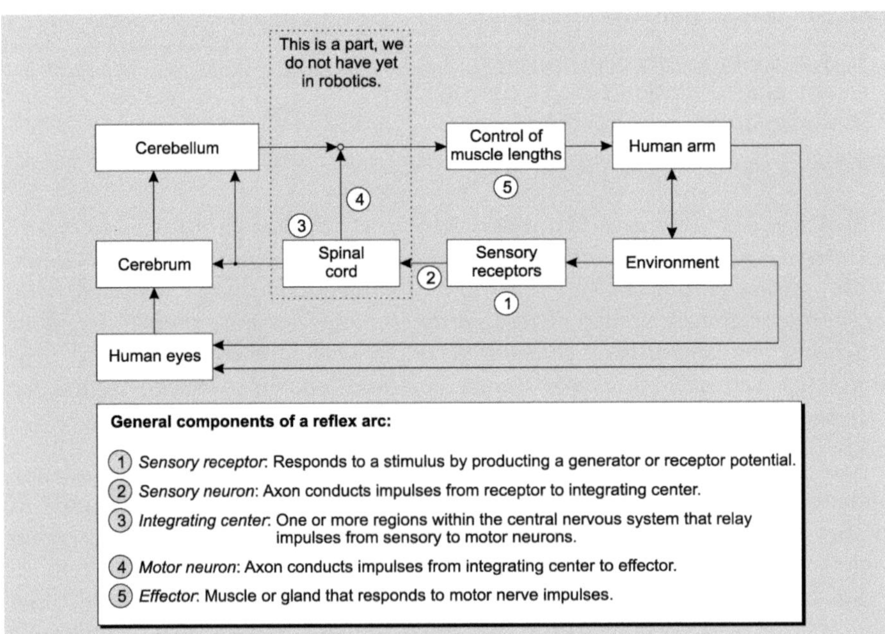

Fig. 1.6 Greatly simplified scheme of human neurophysiological control loops and general components of a reflex arc in accordance with [262].

figure is to transfer the principle of human motor skills (in particular reflexes) to the engineering area.

If we unknowingly touch a very spiky object and suddenly perceive pain, we immediately pull our hand away in a reactive manner (without thinking, that is, without global motion planning). The event of touching the spiky object activates a stimulus to the sensory receptors ① in our skin, which respond by producing a graded generator potential. This potential is conducted to our spinal cord by sensory neurons ②. In the simplest type of reflex, a monosynaptic reflex, the integrating center ③ in our spinal cord is a single synapse between a sensory neuron and a motor neuron. Speaking in the language of an engineer, a synapse is a kind of a switch, comparable to a transistor. More often, the integrating center consists of more than one synapse, which is then called polysynaptic reflex arc. Polysynaptic reflexes enable the involvement of more than one muscle to perform more complex motions, for example, pulling the whole arm, including the hand, away from something. The impulses triggered by the integrating center propagate from the central nervous system along a motor neuron ④ to the part of the body that will respond. This part, a muscle or a gland, is called an effector ⑤, which responds to the motor nerve impulse and performs the resulting movement — the reflex.

Such a synaptical reflex arc can be regarded as a low-level control loop for sensor-guarded motions. Above this loop, we can find further anatomic systems for our motor skills, in particular the cerebrum and the cerebellum. These parts of our brain perceive the stimulus *after* the reflex motion has been initiated and are responsible for higher-level motion planning. To build a bridge from this neurophysiological excursion back to the field of robotics, the human eyes were considered in Fig. 1.6 to have an analogousness to visual servo control.

The quintessence of this excursion is that we do not have a part in the field of robot motion control that is comparable to the human reflex arc. It seems obvious that such a system would make advances in robotics. The whole wide field of sensor-based robot motion control could be reorganized in order to open the gate to new and advanced robotic applications. The major requirement for such a technological advance is that motion control set-points can be generated from any arbitrary state of motion under consideration of dynamic and kinematic constraints. Developing, describing, and evaluating such a module, which is responsible for providing a kind of *robot reflex*, is the aim of this work.

1.3 Outline of This Book

As indicated in Fig. 1.7, this section briefly outlines the rest of this work. To underline the statements of this introductory chapter, Chap. 2 gives a wide overview of the state of the art in technology and the state of the art in

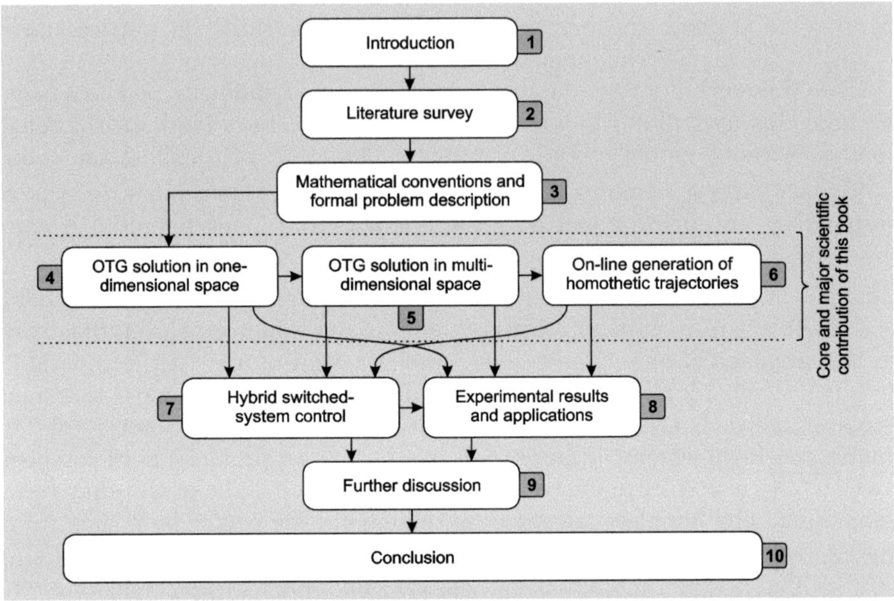

Fig. 1.7 Overview of the coherence of the chapters in this book.

research. Since the actual algorithm for OTG, which will be presented later, is rather large in scale, we describe it strictly mathematically in order to keep this monograph compact. For this purpose, it is necessary to introduce dedicated notations as well as some conventions, which is done in Chap. 3. Based on this, a formal problem formulation for this work as well as a classification of different types of OTG are given. In order to deduce the central algorithm step by step, Chap. 4 delivers the solution for the one-dimensional case, and Chap. 5 transfers this solution to systems with multiple DOFs. One special case, which is practically very important, is the on-line generation of straight-line trajectories; the following chapter extends the existing algorithm to enable the generation of homothetic trajectories.

The central novelty of this work is an open-loop control module, which is able to generate time-optimal and time-synchronized motion trajectories for any mechanical system with multiple DOFs during runtime, that is, on-line during *every* low-level control cycle. This enables new advances for multi-sensor integration in robotic systems, and hybrid switched-control systems become suitable for a wide range of robotic applications. How hybrid switched-systems control can be used for the realization of multi-sensor-based control is addressed in Chap. 7.

Subsequently, simulation and real-world experimental results are discussed in Chap. 8. Besides the exemplary description of several switching events, we describe the OTG interface in relation to higher-level motion planning systems. Furthermore, it is addressed how dynamic system models can be used for the optimization of the resulting trajectory. Before Chap. 10 concludes, Chap. 9 takes up some further marginal topics and deals with them in greater detail.

The Chicago Manual of Style [258] was used as guideline for the writing of this book.

Chapter 2
Literature Survey: Trajectory Generation in and Control of Robotic Systems

This work spans and combines a wide range of research topics. In order to classify this work within the robotics research landscape, this chapter provides a survey about all adjoining fields. Starting with a short section on terminology, the states of the art in technology as well as in robotics research are surveyed.

2.1 Terminology

Before going ahead with the literature survey, some terms and words are briefly defined for the context of this book in order to prevent misunderstandings and misusages as often happens in literature.

Pose/position/orientation
: A *pose* is considered *position and orientation in Euclidian space.* For all other spaces regarded in this book, for example, the joint space or any other multi-dimensional space, we only consider the term *position* for all DOFs. For example, position control in joint space takes all DOFs into account, while position control in Cartesian space only includes the three translational DOFs.

Path planning
: A path is a geometric representation of a plan to move from a start to a target pose. The task of planning is to find a collision-free path among a collection of static and dynamic obstacles. Path planning can also include the consideration of dynamic constraints such as workspace boundaries, maximum velocities, maximum accelerations, and maximum jerks. We distinguish between *on-line* and *off-line* path planning algorithms. Off-line planned paths are static and calculated prior to execution. On-line methods enable the path (re-)calculation and/or adaptation during the robot motion in order to react to and interact with dynamic environments. This means that a robot moves along a path that has not necessarily been computed completely, and which may change during the movement. The

T. Kröger: On-Line Trajectory Generation in Robotic Systems, STAR 58, pp. 11–31.
springerlink.com

term *real-time path planning* is a synonym for *on-line path planning*. It would not make sense to use non-real-time planning algorithms for on-line purposes.

Trajectory planning
A trajectory is more than a path: It also includes velocities, accelerations, and jerks along the path. A common task is to find trajectories for a priori specified paths, which fulfill a certain criterion (e.g., minimum execution time). We distinguish between *on-line* and *off-line* trajectory planning methods. An off-line calculated trajectory cannot be influenced during its execution, while on-line trajectory planning methods can (re-)calculate and/or adapt the robot's motion behavior during the movement. The reasons for this (re-)calculation and/or adaptation can vary: improvement of accuracy, better utilization of currently available dynamics, reaction to and interaction with a dynamic environment, or reaction to (sensor) events. Analogous to *real-time path planning*, the term *real-time trajectory planning* is used in the same way as *on-line trajectory planning*.

Motion planning
Motion planning includes the task of *path planning* and the task of *trajectory planning*. On-line and off-line motion planning is defined analogously to on-line path planning and on-line trajectory planning.

Trajectory generation
This term is used as a synonym for *trajectory planning*.

Motion control
A robotic system interacts with its environment only by moving its kinematic chain. The control of the robot's kinematic chain is meant here. It includes trajectory-following control as well as sensor-guided motion control (e.g., force/torque control).

In particular the words *path* and *trajectory* are not consistently used in literature. The meaning of *off-line* is clear, but *on-line* and *real-time* are often used in different contexts, as will be seen in the two following surveying sections.

2.2 Overview

As depicted in Fig. 1.5 (p. 5), the field of on-line trajectory generation adjoins to (low-level) robot motion control and to (higher-level) robot motion planning, such that it exactly lies in-between these fields. Since these fields belong to the most classical ones in robotics research, we can find plenty of technological and scientific contributions within the last four decades that will be surveyed here.

Fig. 2.1 shows an overview of the treated topics and how this work is related to them. After the reviews in Sec. 2.3 and Sec. 2.4, we will explain in Sec. 2.5 how the work is related to fields depicted in Fig. 2.1.

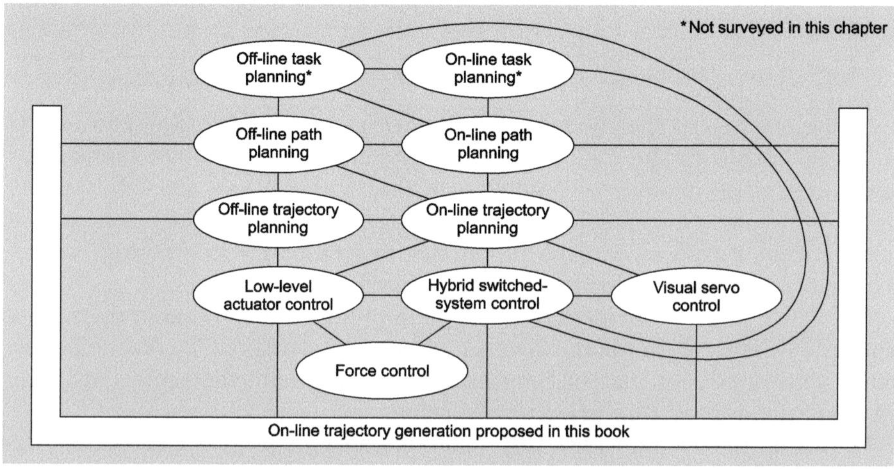

Fig. 2.1 Overview and logical coherence of areas treated in this chapter. The terminology is based on Sec. 2.1.

2.3 State of the Art in Robot Technology

Robot manufacturing companies commonly do not publish detailed information about their control concepts, technologies, methodologies, or philosophies. Hence, this section is very brief, and we predominantly focus on the scientific literature in the next section. To give an overview of the state of the art in robot technology, product specifications of some of the world's leading manufacturers of industrial manipulation robots are surveyed in the following. The text is meant to be neutral and free of bias; manufacturers are named in alphabetical order.

ABB Asea Brown Boveri Ltd. [3] applies a dynamic model in their robot controller called *QuickMove*™and *TrueMove*™concept [1, 2], which enable fast and accurate positioning. Sensor integration is possible through digital and/or analog input and output ports or via field bus systems, but these signals cannot be used for an on-line trajectory adaptation, but for search runs. Even when applying the force control options *RobotWare Machining FC*™or *RobotWare Assembly FC*™[2], the force sensor is not part of the control loop, and only intermediate path segments are adapted based on the sensor signal. Regarding pick and place tasks, ABB offers the *PickMaster*™[4], which uses a computer vision system in order to cope with undefined part poses.

COMAU S.p.A. Robotics [60] offers the *C4G*™robot control unit with two completely different interfaces: 1) The standard *C4G*™[57] interface with the robot programming language *PDL2*™designed for field applications. 2) The open interface *C4G OPEN*™[58, 59] designed for research and development institutions. The programming language *PDL2*™of the standard interface is a high-level Pascal-like language with typical motion planning instructions, but it does not allow for the addressing of a sensor in the feedback control loop. Even the *sensor tracking option* only provides the possibility of modifying a pre-planned trajectory, but the robotic arm remains position controlled. Compared to this, the *C4G OPEN*™interface gives users access to the low-level servo control loops [58, 59]. An external PC communicates with the *C4G OPEN*™controller in cycles of one millisecond and can even set torque set-points, such that the user can absolutely freely develop his own control schemes. This system is excellent for research and development purposes, and users can set up sensor-guided motion control schemes having sensors in the feedback control loop.

FANUC Ltd. [79] provides options for computer vision [77] and force/torque sensor [76] integration for the *R-30iA Mate*™controllers [78]. Nevertheless, these sensors are not part of the feedback control loop; the signals are used for planning a trajectory prior to execution.

Kawasaki Heavy Industries Ltd. [123] offers robot control units named *D-Controller*™[124], which are quite closed and only accessible via the individual and plain *AS-Language*™[125]. Embedding sensors in feedback control loops is not possible in any way.

KUKA Roboter GmbH [153] also applies a dynamic model for the robot motion controller *KR C2*™[152] in order to achieve shorter cycle times and a lower trajectory-following control error. KUKA offers the possibility to embed sensor systems via digital and/or analog input and output ports as well via field bus systems [151]. One interesting system is the *Occubot*™system [150], a robot system for car seat testing. Here, forces and torques are measured in six axes during one single test cycle, and these measured values are used to adapt the trajectory of the next test cycle, such that a desired force can be exerted after a certain number of test cycles.

The robotics division of Mitsubishi Heavy Industries Ltd. [186] provides robot painting systems [185], which are generally not open for any kind of sensor integration. Mitsubishi offers a so-called *open architecture*, which enables customers to interface the robot control unit in time steps of seven milliseconds. Reseller companies such as Battenberg Robotic [18] use this interface for individual add-on controllers.

MOTOMAN Inc. [189] provides the MOTOMAN *NX100*™[190] controller for their industrial robots. Neither the previously mentioned manufacturers nor the *NX100*™of MOTOMAN Inc. provide the feature of embedding a sensor in the feedback control loop [191].

The *Kantana*™manipulator [192] of the Neuronics AG [194] is based on the ideas of using mechanical components of very light weight, of using low-power drives, and of mounting foam around all feather-edged parts of the arm, so that the system can be used in human environments without any further protection mechanisms. The programming interface [193] is very open, such that users can develop their own control schemes, but the standard system does not offer possibilities of integration in the feedback control loop.

Stäubli Faverges SCA [250] offers robot control units, for example, the *CS8C*™controller [248], with digital and/or analog inputs and outputs, which can be used for the retrieval of sensor signals. But as also stated for all other commercial robot manufacturers, it is not possible to put a sensor into the feedback loop of the controller [249]. An interesting option offered by Stäubli is the *Low Level Interface LLI*™[247], which is comparable to the solution *C4G Open*™of Comau Robotics. Pertin et al. describe in [207] how the *LLI*™can be used as an interface for individual control schemes. The *LLI*™accepts position and velocity set-points as well as velocity and torque values for feedforward control.

2.4 State of the Art in Robotics Research

Fig. 2.2 indicates a first brief classification of this work with regard to adjoining fields, which are surveyed in the following three subsections.

General overviews of *all* fields can be found in some robotic textbooks. The Springer Handbook of Robotics edited by Siciliano and Khatib [240] includes a chapter about off-line motion planning written by Kavraki and LaValle [122] and a chapter about on-line motion planning written by Brock, Kuffner, and Xiao [36]. A very recent and also very excellent textbook on on-line (but mostly off-line) trajectory generation concepts was written by Biagiotti and Melchiorri [24]; in particular the part [25] contains fundamental methods that are also applied in this work. Basic concepts of robot motion control were formulated by Chung, Fu, and Hsu [55]. Shorter summaries are given in the textbooks of Craig [62], of Fu, Gonzalez, and Lee [94], and of Spong, Hutchinson, and Vidyasager [246]. In general, it is of course a voluminous task to classify the whole scientific literature in this field, because we can find contributions from four decades ago and almost every work considers different assumptions and basic conditions, but this classification has to be done in order to place this work correctly within this wide field of literature.

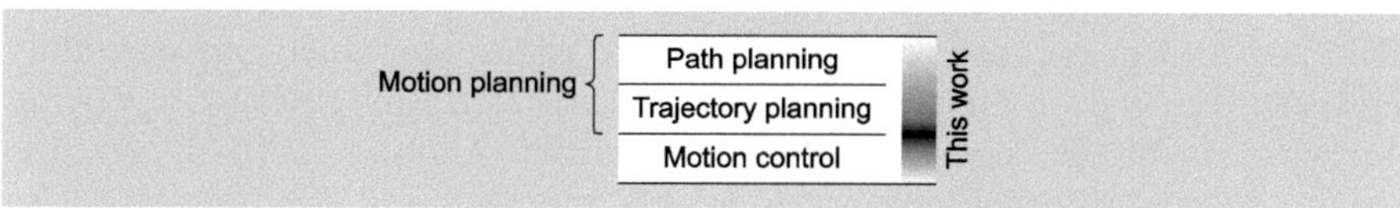

Fig. 2.2 Brief classification of this survey based on the terminology of Sec. 2.1.

2.4.1 Path Planning

Off-Line Path Planning Methods

Off-line path planning methods are not directly related to this work, but as on-line path planning concepts are often based on off-line ones, we refer to a selection of surveys and books: Lozano-Pérez [169], Lindemann and LaValle [166], González-Baños, Hsu, and Latombe [101], and LaValle [156].

On-Line Path Planning Methods

On-line path planning is generally only reasonable if paths *and* trajectories are considered. Hence, we talk about *on-line motion planning* or — to use a new eligible term for this field — real-time adaptive motion planning (RAMP, [266]). This field of research is very young and publications are rare. There are many works assuming dynamic but exactly known environments, which is not meant here. We assume robots that have to act in a dynamic and/or unknown environment, and which are equipped with sensor systems to react to and interact on-line with (unknown) static or dynamic obstacles, events, or (abrupt) changes of task parameters. As a result, RAMP takes place in the *configuration×time* (CT) space.

The research groups of Brock and Xiao focus on real-time capable methods for (multi-)robot motion planning. The use of splines is one very common method to represent calculated trajectories; it is the task of a motion planning algorithm to calculate respective sets of spline knots and trajectory parameters during runtime. [266] proposes an on-line motion planning approach; paths and trajectories are calculated on-line in CT-space, such that the system can act in unknown dynamic environments. Yang and Brock [280] use collision-free vertices ("milestones") and edges on a roadmap, which is another kind of representation, to represent currently planned trajectories. These knots or milestones are generated from an overall view onto the robotic system and its environment — it is in particular a kind of motion planning from a global point of view. The PhD thesis of Brock from 1999 [33] as well as follow-up works published together with Khatib and Kavraki [34, 35] introduced the elastic strip framework. The basic idea of this work is to locally deform a previously planned trajectory in order to avoid moving obstacles inside a collision-free "elastic tunnel", such that the robot can move task-consistently from the initial to the desired target pose in a three-dimensional workspace.

Jaillet and Siméon [118] on the contrary use rapidly exploring random trees (RRT) as a local planner to update a global roadmap, which was originally represented by a probabilistic road map (PRM). This way, arbitrarily moving obstacles can be also considered during runtime. Another method, which is based on a potential field approach, avoids unknown and dynamic obstacles and was presented by Ögren, Egerstedt, and Hu [195]. In the work of Mbede et al. [177], a neuro-fuzzy controller is used to locally modify the base motion of a mobile manipulator in order to avoid a moving obstacle. Based on the contribution of Mbede et al., Merchán-Cruz and Morris [180] published a work about motion planning for cooperating manipulators in 2006. Li and Latombe [161, 162] presented works on an on-line planner for the special purpose of planning the motions of two robot arms taking arbitrarily positioned parts from a conveyer belt.

2.4.2 Trajectory Planning Concepts

Off-Line Trajectory Planning Methods

Even if off-line trajectory planning is not a subject in this work, many on-line trajectory planning concepts are rooted in ideas from off-line concepts. Roth and Kahn [120] belong to the pioneers in the field of time-optimal trajectory planning. In 1971, they published methods of optimal, linear control theory and achieved a near-time-optimal solution for linearized manipulators. The resulting trajectories are jerk-limited and lead to smaller trajectory-following errors and to less excitation of structural natural frequencies in the system.

In 1982, the work of Brady [29] introduced several techniques of trajectory planning in joint space and Paul [203, 204] and Taylor [256] published works about the planning of trajectories in Cartesian space in parallel to Brady. Lin, Chang, and Luh [165] published another purely kinematic approach in 1983 as did Castain and Paul [43] in 1984 and Chand and Doty [44] in 1985.

In 1984, one of the early publications of Hollerbach [114] first introduced the consideration of the nonlinear inverse robot dynamics for the generation of manipulator trajectories. The aim here is to exhaust the maximum actuator forces and/or torques as well as possible in order to achieve shorter execution times. The basic idea is to represent the path as well as the trajectory by a set of parametric functions, which are employed in the dynamic model of the manipulator. This way, the optimization problem can be described for an arbitrary number of DOFs. Later works of Hollerbach's group are based on this idea [10, 11, 223].

During the middle of the 1980s, three groups developed techniques for time-optimal trajectory planning for arbitrarily specified paths: Bobrow, Dubowsky, and Gibson [27, 28], Shin and McKay [238, 239], and Pfeiffer and Johanni [208]. Based on the approach of Hollerbach, that is, describing the robot dynamics of dependence on a parametric path representation, a maximum acceleration can be calculated for each point of the trajectory. These maximum acceleration values correspond to maximum actuation forces and/or torques. Furthermore, maximum actuator velocities are taken into account, such that a characteristic curve for the maximum velocity can be calculated. This so-called maximum velocity curve describes the maximum velocity for each trajectory point and has to be regarded during the trajectory planning phase, that is, the time-optimal trajectory must avoid crossing the limit curve in order to minimize execution time. A time-optimal trajectory can be found by determining the switching points between positive and negative maximum acceleration values, which have been calculated beforehand. The algorithms of the groups around Bobrow, Pfeiffer, and Shin differ in their way to find these switching points.

Independently from these three groups, Rajan [218] presented a spline-based approach for minimum-time *motion* planning (cf. Fig. 2.2). Here an initial and a goal position are given in configuration space; path *and* trajectory are calculated and represented by cubic splines. Geering, Guzzella, Hepner, and Onder [99] discuss the same topic and extend it to different types of robot kinematics.

Shiller and Dubowski [235] used *B-splines* as extension of the basic algorithm of Bobrow, Pfeiffer, and Shin as did Takayama and Kano [254] later. A B-spline curve is an extended version of Bézier curves that consists of segments, each of which can be viewed as an individual Bézier curve with some additions; a detailed introduction can be found in [212].

In 1988, Kyriakopoulos and Saridis [154] extended the basic trajectory planning approach of Bobrow/Pfeiffer/Shin by introducing a minimum-jerk criterion in order to achieve a better trajectory-following behavior. Tan and Potts [255] first transferred the ideas of Bobrow, Pfeiffer, and Shin to a discretely working system that was based on a discrete dynamic model. One year later, in 1989, Slotine and Yang [244] proposed an alternative algorithm for the works of Bobrow, Pfeiffer, and Shin. To abandon the computationally expensive calculation of the maximum velocity curve, a method for the analytical calculation of a limit curve is proposed, and conditions for characteristic switching points on this curve are determined in order to achieve a time-optimal trajectory. In the same year, Olomski [196] presented detailed experimental results and used the off-line generated trajectories as input values for different trajectory-following control schemes.

At the same time, Chen and Desrochers [49, 50, 51] proved that, as long as only dynamic constraints are considered, a time-optimal trajectory can only be achieved if at least one actuator permanently runs in saturation. McCarthy and Bobrow [178] present a similar proof but for the more general case of robots with arbitrary amounts of DOFs.

At the beginning of the 1990s, Shiller and Lu [236, 237] extended the algorithm of Bobrow, Pfeiffer, and Shin and considered dynamic singularities. A dynamic singularity is a part of the trajectory, at which at least one actuator does not contribute to the acceleration along the path.

Apart from the works mentioned above, Simon and Isik [243] presented a concept using trigonometric splines in 1993. The basic idea is to connect knots in joint space that have previously been calculated by the inverse kinematic model, with parameterized trigonometric functions. Some of the parameters can be chosen to minimize the jerk, which results in a very smooth and harmonic motion in joint space.

In 1994, Dahl [63] improved the basic algorithm of Shiller and Lu, such that it becomes more compact. Furthermore, Dahl presented more advanced experimental results. Two years later, Fiorini and Shiller [89] presented a further off-line algorithm for known dynamic environments. It is based on the assumption that the environment is completely known beforehand, that is, all static objects are known, and all moving objects are known with known trajectories.

Also in 1994, von Stryk and Schlemmer [252] evolved experimental results for minimum-time and minimum-energy trajectories for a Manutec r3 industrial robot. The research group of von Stryk published further works on trajectory optimization methods, for example, [109]. In 1996, Žlajpah [270] slightly extended the basic approach of Bobrow, Pfeiffer, and Shin by embedding task constraints. The original algorithm becomes redefined and different areas below the maximum velocity curve are defined and taken into account during the calculation of the switching points.

Based on Lie groups and Riemannian geometry, which were applied together in research on robotics dynamics by Park, Bobrow, and Ploen in 1995 [201], Žefran, Kumar, and Croke [269] suggested an off-line method for the generation of task space trajectories from an initial to a target pose (both with zero velocities); this is an approach that considers multiple DOFs. The key contribution of this work is a measure for the smoothness of a *multi-dimensional* trajectory that is supposed to be optimized by the proposed algorithm. Furthermore, dynamic system constraints such as boundary velocity and acceleration curves are considered.

A method for the generation of minimum-jerk trajectories, but without consideration of robot-dynamics, was contributed by Piazzi and Visioli [209, 210] in 1998 and 2000. In a follow-up work from 2002, Bianco and Piazzi embedded robot dynamics into this approach [26].

A general overview of basic off-line trajectory planning concepts is presented in the textbook of Khalil and Dombre [126]. A more recent approach of Lambrechts, Boerlage, and Steinbuch from 2004 suggests the generation of very smooth trajectories, whose jerk derivatives are limited as well [155], but there, only one DOF-systems are considered, and the method requires initial and goal velocities of zero.

On-Line Trajectory Planning Methods

An on-line modification of a planned trajectory may have several reasons: 1) The trajectory becomes adapted in order to improve the accuracy with a path specified beforehand; 2) The robotic system reacts on sensor signals and/or events that cannot be predicted beforehand, because the robot acts in a (partly) unknown and dynamic environment. These two reasons are rooted in the field of robotics, but there is another field, in which on-line trajectory planning plays an important role: CNC machine tools. The following three paragraphs survey these fields.

Improving Path Accuracy

All previously described off-line trajectory planning methods assume a dynamic model that describes the behavior of the real robot exactly. In practice, this is often not the case, and some robot parameters are only estimated, some dynamic effects remain unmodeled, and system parameters may change during operation. If this is the case, the resulting robot motion is not time-optimal anymore and/or the maximum actuator forces and/or torques are exceeded, which leads to an undesired difference between the specified and the executed path. In the following, we give a survey of methods that adapt the trajectory on-line in order to improve the path accuracy.

In 1989, Dahl and Nielsen [64] suggested an on-line trajectory adaptation method. The approach is based on the basic algorithm of Bobrow/Pfeiffer/Shin, and adapts the acceleration along the path, and furthermore the parameters of the trajectory-following controller are adapted on-line, such that the underlying trajectory-following controller becomes adapted, depending on the current state of motion.

Independently from Dahl, van Aken and van Brussel [265] proposed a concept that uses one-dimensional parameterized acceleration profiles along the path in joint space instead of adapted splines. These profiles are computed on-line under consideration of the dynamic model of the manipulator.

Also independently from the two latter works, Bestaoui [23] proposed another algorithm for on-line trajectory generation at the same time. This method replans the trajectory based on state-dependent acceleration and velocity constraints if an intermediate trajectory knot is attained, or if the difference between the real and the desired trajectory is greater than a certain threshold.

The approaches of Cao, Dodds, and Irwin from 1994 [41] and 1998 [42] use cubic splines to generate smooth paths in joint space with time-optimal trajectories. A cost function is used to define an optimization problem that considers the execution time and the smoothness of the path. The cost-function can be minimized during the motion by applying the efficient numerical quasi-Newton method proposed by Davidon, Fletcher, and Powell [65, 92]. Thus, new intermediate knots are calculated. Robot dynamics are only considered by a set of parameters that must be determined empirically.

The works of Bazaz and Tondu [20, 21] (1997, 1999) extend the work of Bestaoui [23] and use segments of cubic splines to interconnect trajectory segments instead of fourth-order polynomials to describe the trajectory.

In 1998, Schlemmer and Gruebel [224] suggested an on-line method that embeds a static environmental model, such that the optimization problem becomes extended. Not only the minimum time criterion is applied but also a further function representing the distance between the robot and its environment is taken into account. Hence, the aim is to find a compromise trajectory represented by cubic splines that minimizes the execution time and that maximizes the distance to the environment.

Constantinescu and Croft [61] suggest a further improvement to the approach of Shiller and Lu [234, 236, 237] in the year 2000. The additional limitation of the derivative of actuator forces/torques leads to a limitation of the jerk in joint space and thus to smoother trajectories with better following behavior.

The PhD thesis of Pietsch [214] from 2003 is based on the method of Dahl and Nielsen and improved the concept by combining adaptive control and time-optimal trajectory planning.

In the same year, Macfarlane and Croft presented a jerk-bounded, near-time-optimal, one-dimensional trajectory planner in [174] that uses quintic splines that are computed on-line. Similar to the approach of Andersson [13, 14] from 1988, which will be described in the next paragraph, Macfarlane and Croft use quintic polynomials. As an extension to Andersson, they proposed an additional algorithm, which guarantees that these polynomials will perform a limited jerk along the path. Experimental results underline the usability of this approach, but robot dynamics are not considered.

Together with Owen and Benhabib, Croft publishes a further work on on-line trajectory planning [199, 200]. Here, a cooperating robot task is considered. An off-line planned trajectory becomes adapted on-line in order to respond to unmodeled disturbances and to maintain the desired path.

A work of Kim et al. from 2007 [130] implies improving the algorithm of Macfarlane and Croft [174] by taking robot dynamics into account, but only simulation results are shown.

Sensor-Based Trajectory Adaptation

The last paragraph presented an overview of on-line trajectory generation methods for improving the path accuracy, while this one focuses on the on-line consideration of sensor signals.

The works of Andersson [13, 14] from 1988 and 1989 present a Ping-Pong-playing PUMA 260 manipulator, whose trajectory is generated on the basis of a three-dimensional stereo-vision system. The position progression of a trajectory is represented by quintic polynomials that are parameterized by Andersson's algorithm. One particular property is that the algorithm works with arbitrary initial and goal conditions, in particular target velocities unequal to zero.

In 1993, Lloyd and Hayward [168] proposed a technique to perform transitions between two different path segments. On the one hand, the manipulator dynamics are considered based on the basic trajectory generation approach of Bobrow/Pfeiffer/Shin [27, 28, 208, 238, 239], and on the other hand the transition window technique first proposed by Paul [205] and Taylor [256] is applied. The two path segments, which are to be blended together within the transition window, need not be known in advance. The trajectory-describing blend function for the transition window is calculated on-line and considers acceleration and velocity bounds.

Inspired by human arm movements and based on the works of Lloyd and Hayward, the research group of Flash [98, 222] published further works in the same field in 1999 and 2001. Their superposition principle incorporates position, velocity, and acceleration constraints. As in the method of Lloyd and Hayward, the target position in joint space can change arbitrarily, and within a certain superposition window, the current trajectory "fades out" while the new one "fades in."

In 2002, Liu [167] presented a one-dimensional approach that calculates linear acceleration progressions on-line by parameterizing the classic seven-segment acceleration profile [43]. This way, the system can react to arbitrary target position switchings at any time. Liu's approach is strongly related to a part of this work, but as will be described in Chap. 4, it is incomplete and erroneous.

Two years later, in 2004, Ahn, Chung, and Youm [5] proposed a work for the on-line calculation of one-dimensional trajectories for any given state of motion *and* with arbitrary target states of motion, that is, with target velocities and target accelerations unequal to zero. Sixth-order polynomials are used to represent the trajectory, which is called arbitrary states polynomial-like trajectory (ASPOT). The major drawback of this work is that neither kinematic nor dynamic constraints can be regarded for on-line trajectory generation, such that this method cannot be applied in practice.

In a work from 2005, Chwa, Kang, and Choi [56] presented an advanced visual servo control system, whose purpose is to intercept fast and arbitrarily-moving objects with a robot. The proposed on-line trajectory planner considers the system dynamics of a two-link planar robot and works in two DOFs. A drawback of this work is its missing generality, and only simulation results are shown.

The title of the work of Kang [121] implies the description of an on-line trajectory planning method, but only basic and well-known kinematic and dynamic principles of robotics are presented.

Broquère, Sidobre, and Herrera-Aguilar [38] published a work in 2008 that uses an on-line trajectory generator for an arbitrary number of independently acting DOFs. The approach is very similar to the one of Liu [167] from 2002 and is based on the classic seven-segment acceleration profile [43], but this approach is unfortunately also incomplete (cf. Chap. 4) and can only perform reactions if the current acceleration value of a DOF is zero. An additional very recent work of Haschke, Weitnauer, and Ritter [108] presents an on-line trajectory planner in the very same sense as this book does. The proposed algorithm generates jerk-limited trajectories from arbitrary states of motion, but it suffers from numerical stability problems and only allows target velocities of zero.

On-Line Trajectory Generation for CNC Machine Tools

Besides the field of robotics research, we can also find approaches for trajectory generation in the field of CNC, that is, machines reading G-code instructions and driving machine tools. G-code is the numerical control programming language. We neither give an overview of CNC systems nor introduce this field here. Therefore, please refer to basic textbooks, for example, the one of Suh, Kang, Shung, and Stroud [253]. Another brief survey of the field of machine tools is given in [133]. In the following, only recent CNC-related works that address the field of on-line trajectory generation are briefly surveyed.

Non-uniform, rational B-splines (NURBS) are commonly used as representation of free-form shapes in the field of CAD/CAM systems (computer aided design/computer aided manufacturing). During the process of machining, these free-form shapes are comparable to arbitrarily specified paths in the field of robotics, and — hence — the basic task of accurate path-following, is very similar. In the field of CNC, the path is always specified, and we only have to work on interpolators. On-line interpolators, which can also be regarded as on-line trajectory generators, are the focus of current research in machine tools. An introduction to NURBS can be found, for example, in the textbook of Piegel [212] or in a short survey by him in [211].

Works on on-line NURBS interpolation [48, 52, 54, 132, 157] usually consider the complete system dynamics for setting up CNC control units, and since the path is completely known in advance, the dynamics can also be calculated beforehand, but this way, unforeseen and unmodeled effects are excluded, and the responsibility of proper functioning is passed to the tracking controllers. In order to support these controllers, it is necessary to adapt the desired trajectory on-line in order to prevent them from bringing actuators to saturation and therewith increasing the tracking error. All mentioned real-time NURBS interpolation techniques are suited for a priori-specified paths represented by NURBS, and hence these works are related to the survey about on-line trajectory generation for improving the path accuracy (p. 19).

Besides these on-line methods, we can also find recent works in the field of CNC machine tools that transfer basic (off-line) methods from the field of robotics. Dong, Ferreira, and Stori [69] present an off-line algorithm to calculate a varying feedrate for CNC machine tools in order to achieve a time-optimal trajectory. The approach from 2007 uses the algorithm of Bobrow/Pfeiffer/Shin [27, 28, 208, 238, 239] as the basis for calculating a feedrate progression for a given path.

2.4.3 Robot Motion Control

Regarding Fig. 2.2 (p. 15), the field of robot motion control is the third field besides path and trajectory planning that relates to this book. Although the latter two subsections focused on robotic manipulators, we address a wide field of robotic systems in general and hence do not focus specifically on any concrete motion control scheme or any concrete kinematic. Only brief overviews of robot motion control in general and hybrid switched-system control are given in the following.

Actuator/Joint Control in Robotics

The overview of low-level robot motion control schemes in the Springer Handbook of Robotics [55] was already mentioned earlier. Additional textbooks describing this subject are by de Wit, Siciliano, and Bastin [278], by Kozłowski

[135], by Sciavicco and Siciliano [229], and by Khalil and Dombre [127]. Many of the described approaches are based on the early works of Whitney [274] and Paul [206]. Another important issue is embedding rigid body dynamics to set up a dynamic model of the robot. The works of Featherstone [81, 82, 83] are to be mentioned as fundamental to this field.

In the following part, we will focus on hybrid switched-system control, which constitutes the next control layer above the actuator controller. For a hybrid switched-system controller, it is essential that the underlying actuator control scheme is stable, that is, if we cannot prove stability for the inner control loops, stability for the outer loops will be even harder to prove.

Hybrid Switched-System Control

Hybrid switched-control systems are systems that comprise a family of continuously working subsystems and a supervisory discrete system that switches between them. Fig. 2.3 shows an abstract and simple scheme of a hybrid switched-system control architecture for a one-DOF system, that is, with one actuator only (e.g., a simple linear positioning unit). This system can be controlled on the basis of different sensor signals (distance, vision, and force/torque), and depending on the current task and/or depending on the current system state and/or depending on time, one of the controllers can be selected. The force/torque, the distance, and the visual servo controllers of Fig. 2.3 are closed-loop controllers; the velocity and the position controller are open-loop controllers (i.e., on-line trajectory generators) as they will be proposed in this work. The difference between these latter two modules is

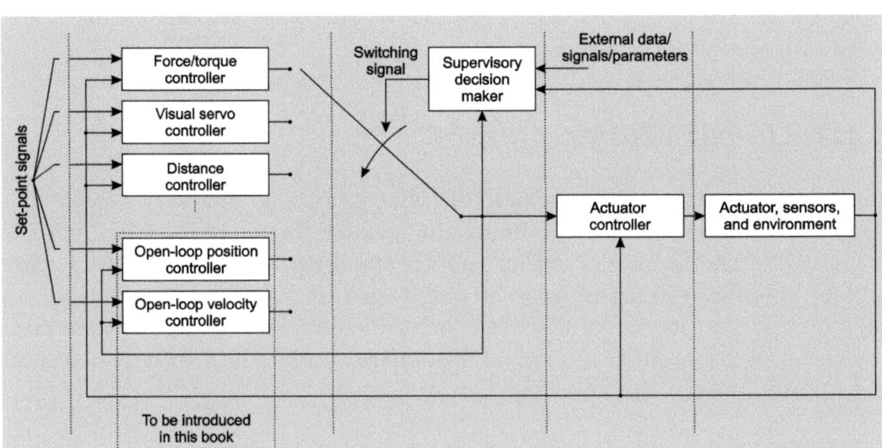

Fig. 2.3 Abstract control scheme of a hybrid switched-system for one single DOF, that is, one single actuator. The dotted vertical lines will be explained in Sec. 2.5 (p. 30).

that the velocity controller does not consider target positions and only leads a single DOF to a certain velocity under given acceleration and/or jerk (additionally also the derivative of the jerk) constraints. Therefore, it is of a much simpler manner than the open-loop position controller. The system depicted in Fig. 2.3 can, of course, be extended to multiple DOFs, as will be shown in Chap. 7 (p. 105).

When we take a multi-DOF robotic manipulation system with many different sensors for granted, and when we consider a system that enables the execution of sensor-guided motion control commands in any DOF, it becomes self-evident that we need to switch discretely between several continuously working (open- and/or closed-loop) controllers at any time. Hence the analysis of hybrid switched-system control is one fundamental part of the work presented here.

Especially the works of Branicky [30, 31] and Liberzon [163, 164] provide elementary concepts to develop and analyze hybrid switched-system control techniques. In particular the stability analysis is of fundamental interest here, because the stability of a switched-system cannot be assured by the stability of each single sub-controller. To prove the stability of hybrid switched-systems can be extremely difficult and many researchers are working on analyzing such stability questions. It may happen that a set of stable subsystems becomes unstable if the switching between them occurs inappropriately [32, 164, 277]. In the field of stability analysis, we can distinguish between techniques for linear [66, 164, 181] and nonlinear [67, 179, 279] switching systems.

The discrete switching system by itself can of course not work without the respective continuously working subsystems. Hence, we also refer to basic works for the two most relevant candidates: force/torque control and visual servo control. Distance control (cf. Fig. 2.3) usually works in one DOF only and is a trivial task.

Force/Torque Control

Robotic applications such as grinding, assembly, or deburring involve an extensive contact between the robot and its environment. In order to establish this contact, force/torque control is applied. In all previously mentioned works in this section, only pose and/or position control was regarded, and this is the first scenario, in which a further sensor is part of the feedback control loop.[1] These sensors can be either wrist-mounted force/torque sensors (e.g., [15, 119]) or joint torque sensors. The latter method was, for example, successfully realized with the DLR light-weight arms [8, 149], which possess joint-integrated torque sensors. Force/torque control, based on sensor signals from wrist-mounted sensors, can be applied to a much wider range of robot systems, since only actuator position feedback — as it is common for

[1] Force control concepts can be distinguished in active and passive methods. Passive ones do not consider a force/torque sensor in the feedback control loop [276] and are excluded here.

industrial robots — is required. The following paragraphs give a brief overview of this field.

Whitney [275] and Mason [176] belong to the pioneers in the field of compliant motion control. Based on their work, numerous approaches have been published in this field. Khatib [129] formulated his key work about the *operational space approach*, and especially the group of De Schutter [226, 227, 228] contributed promising concepts to the community and coined the term *Task Frame Formalism* [40], which enables the development of compliant motion solutions on an abstract programming level.

These works understand force/torque control as one basic element of compliant motion concepts [267]. Three different approaches are known from literature: 1. Impedance control [113], which uses relationships between acting forces/torques and manipulator pose to adjust the mechanical impedance of the end-effector to external forces/torques; 2. Parallel control [53], which enables to control both, force/torque and pose, along the same task space direction. 3. Hybrid force/torque and pose control, which controls force/torque and pose in two orthogonal subspaces [217]. To realize this approach, we have to pay attention to the problem of orthogonality as stated by Duffy [72], who extended the approach such that it is consistent, independent of units, and independent of any origin coordinate system.

General overviews about the field of force/torque control can be found in the Springer Handbook of Robotics [267] and, for example, in the textbooks of Siciliano and Villani [241] or Gorinevsky, Formalsky, and Schneider [102], respectively. The previously mentioned textbooks of de Wit et al. [278], Craig [62], Khalil et al. [128], Kozłowski [135], Sciavicco et al. [229], and Spong et al. [246] also give brief surveys of this field. Another good overview of this field can be found in the work of Vukobratović and Šurdilović [268].

The PhD thesis of Reisinger, who belongs to the same research group as the author does, contains a control framework for the contact transition of force/torque-controlled parallel kinematic machines [220].

Visual Servo Control

If computer vision data is used for robot motion control, we speak about *visual servo control*. Together with force/torque control, powerful robotic systems can be achieved, because — similar to human beings — robots can use two very complementary sensors, one to recognize the global task environment, and one for fine (contact) motions. A camera may be mounted directly on a robot (eye-in-hand) or the camera can be fixed somewhere in the robot's workcell, such that it observes the robot motion from a stationary pose. The field of visual servo control consists of three domains: low-level image processing, computer vision, and control theory. Here, only a brief overview with regard to hybrid switched-system control is given.

Basic overviews were presented by Chaumette and Hutchinson [45, 46, 47], which are regarded as a very good introduction for control engineers, who often are not familiar with the field of computer vision. The origins for most works on visual servo control can be found in the publications of Weiss, Sanderson, and Neuman [273] and of Feddema and Mitchell [84]. Detailed introductions to image processing and computer vision algorithms can be found, for example, in the textbooks of Wahl [271], by Forsyth [93], and of Ma, Soatto, Košecká, and Sastry [171].

Regarding Fig. 2.3, visual servo control is supposed to be applied as one continuously working submodule in a hybrid switched-system. Baeten and De Schutter [16] present a very comprehensive work, in which computer vision and force/torque control become unified in the Task Frame Formalism [40]. Another more recent approach of Gans and Hutchinson [95, 96, 97] suggests two visual servo controllers as submodules in a hybrid switched-system. Assuming an eye-in-hand camera setup, the first control module uses the camera position to calculate an error signal in the feedback loop of the control law and the second submodule uses image features. Stability is proven by means of a state-based switching scheme.

2.4.4 Human-Inspired Motion Analysis

When considering reflexes and human motion patterns as inspiration for the development of robotic systems, as described in the explanation of human neurophysiology (Sec. 1.2, p. 6), we also have to compare both worlds with each other, and we should try to build a bridge between both fields. Flash and Hogan [91] present a mathematical model based on human arm movements. A set of tasks (including compliant motion tasks) was used to observe trajectories taken by a human arm. The observed trajectories, in particular the magnitude of the jerk integrated over the entire movement, were used to rate the performance of an executed trajectory. In a later work of Flash [90], it is analyzed what human arm trajectories would look like if test candidates performed an aimed arm movement to a visually recognized target location, and if this target location suddenly changed. Even if it is not clear how this motor task is performed, it could be stated that after a new visual stimulus appears, an old "plan" is aborted and a new one appears during the movement. A further work of Henis and Flash [110] underlines this book; here time intervals between a visual stimulus and the respective trajectory modification are investigated. Depending on the current trajectory, that is, the current state of motion with regard to the old and the new target, such an interval takes between 10 and 300 milliseconds. Till today, it is not clear how our neurophysiological system exactly works. Human motion patterns are obviously generated on different levels. A monosynaptic reflex, as described in Sec. 1.2 (p. 6), constitutes the lowest level. The motion patterns are furthermore represented in multiple layers in the human motor cortex of the cerebrum (cf. Fig. 1.6, p. 6), which contains regions that are involved in the planning, control, and execution of human motor skills [232, 263, 272].

We can also find works that consider human-like reflexes in the field of robotics, in particular in the fields of humanoid, biped, or quadruped robots, in which common motion pattern generators are employed to generate set-points for low-level controllers that are responsible for standing upright. After a (sensor) event, this pattern can be adapted by adding a further previously *learned pattern* in order to assure that the system stands upright. Zaier et al. [282, 283] propose such a control scheme for humanoid robots. The learned pattern, which is added to the currently generated one, always fades smoothly in and out, in order to prevent jerky motions. The same idea was applied to prosthetics in [230].

In [106], Haddadin, Albu-Schäffer, De Luca, and Hirzinger present a very impressive work on the detection of unforeseen collisions and respective reaction concepts. Based on previous works of the authors [103, 104, 105, 170], five different collision (= sensor event) reaction strategies are investigated[2]: **1)** The robot shows no reaction at all and continues to follow the reference trajectory; **2)** The robot is stopped as soon as a collision is detected. This is obtained by using the actual joint position, which was measured at the time instant of collision detection, as set-point for the position controller; **3)** Switch from position control to zero-gravity torque control [6, 7], letting the robot behave in a very compliant way; **4)** Switch to torque control with gravity compensation but, in contrast to 3), use joint torque feedback and the signal of the estimated external torque, which is used as a collision signal, to scale down both the motor inertia as well as the link inertia, thus obtaining an even "lighter" robot; **5)** Use the estimated external torque to implement an admittance controller. By defining the desired velocity in the opposite direction of the external torque estimation, the robot "flees" from this disturbance. The strategies 3 – 5 contain switchings from trajectory-following control to sensor-guided robot motion control; this monograph considers the opposite way of switching: from sensor-guided motion back to trajectory-following control (cf. Fig. 1.4, p. 4).

2.4.5 Own Works

This subsection shall briefly classify a selection of previous works of the author with regard to the previously surveyed fields of robotics research. Starting with works on force/torque control for robotic manipulation systems, in particular adaptive control and model-following control schemes [87, 136, 198, 260], especially works on practical implementations of the Task Frame Formalism [137, 138, 139] were done by the author. Another field addresses the fusion of force/torque and acceleration signals in order

[2] The rest of this paragraph is written in accordance with [106].

to separate contact forces/torques and forces/torques caused by inertial effects [143, 144, 145] with the aim to improve force/torque control quality in practical applications [147, 148]. [88] introduces *Manipulation Primitives* as an interface to hybrid switched-systems applied in the field of robot motion control. Besides these objectives, software engineering and real-time aspects commonly play a fundamental role for practical and elegant implementations of the mentioned techniques; over the years, a framework based on a dedicated real-time middleware was conceived and finally presented in [86]. Besides these basic works, experimental achievements in these fields were presented in [140, 141, 142]. The last work to be mentioned here is [146], which is closely related to this book and presents a preliminary stage of on-line trajectory generation in hybrid switched-systems.

2.5 Conclusions and Classification of This Work

After this broad and, in some areas, thorough survey, we are now able to concretize the motivation for this work and to classify it within the robotics research landscape.

The majority of the surveyed concepts for off-line and also on-line motion generation produce a motion along a specified path. But, is this a good approach? — For purely position/pose and/or trajectory-following controlled motions: Sure and without restriction of any kind! But: When we execute sensor-guided motions, for example, by force/torque or by visual servo control, we do not have a predefined path anyway, because the robot motion directly depends on the sensor signal. We have to dismiss the path during sensor-based motion control! As soon as we embed sensor-guided or sensor-guarded motions, there is no predefined path anymore. In particular, we have to say good-bye to trajectory planning and reference trajectories along previously specified paths. There is no path that can be exactly followed, because everything may depend on sensors whose signals cannot be foreseen.

For further advancements in the field of robotics, sensor integration is absolutely indispensable! We can find plenty of approaches in the scientific arena, but when looking at the state of the art in technology (Sec. 2.3, p. 13), almost none of these approaches can be found. The author believes that there is one missing part in the robot control architecture as indicated in Fig. 1.4 (p. 4) that enables switchings from sensor-guided to trajectory-following operations. Furthermore, commercial robot manufacturers have to develop reliable and deterministically working machines. These companies need concepts to react safely to sensor malfunctions. How should a robot behave if a sensor stops delivering a signal during a high-speed sensor-guided motion? If a robot performs a sensor-guided motion in all of its DOFs (e.g., zero-force-control during a teach-in process) how can the controller — except when performing an emergency stop — take over the guidance of the robot if something unforeseen happens?

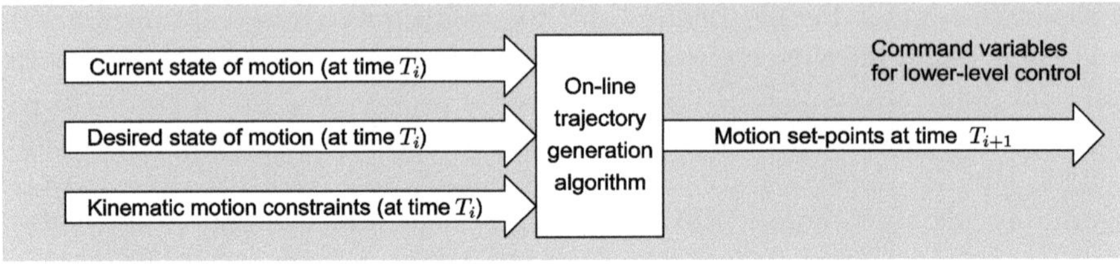

Fig. 2.4 Input and output values of the on-line trajectory generation algorithm.

These are simple questions that are mostly not relevant for research institutions. On the one hand, we (researchers) often complain about inaccessible commercial control units, and that sensor-guided motions are not possible, but on the other hand, we also do not have a general concept that is able to generate set-points for a trajectory from arbitrary initial states of motion to arbitrary target states of motion under consideration of kinematic and dynamic constraints — as would be required for practical realizations.

One of the goals of the author is to contribute to the solutions of these problems. The concept proposed in this monograph does not deliver a complete solution — it is just one further piece of technology to be developed with this work. To place this work in the wide landscape of robotics research, the author suggests considering two different points of view:

Horizontal view
Here, we take a model of multiple horizontal layers for robot motion control into account. As depicted in Fig. 1.5 (p. 5), on-line trajectory generation is considered together with sensor-based motion control as one horizontal layer in an abstract robot control model. This layer constitutes the interface between higher-level (on-line and/or off-line) task and motion planning and lower-level actuator control. This point of view seems to be common in the field of computer science.

Vertical view
Here, we consider vertical layers of a hybrid switched-system control scheme as shown in Fig. 2.3 (p. 24), which shows such a simple scheme and indicates the layered structure via the dotted lines. The on-line trajectory generation algorithm is regarded as one continuously working submodule that is able to take over control from any arbitrary state of motion and to transfer the system to a desired state of motion in the shortest possible time. This figure resembles the view of control engineers.

The aim of this work is to develop an algorithm for time-discrete systems as depicted in Fig. 2.4. This on-line trajectory generation algorithm can also be regarded as feedforward or open-loop controller. One can also consider it as a feedback on the current state. At a time instant T_i the algorithm receives the parameters of the current state of motion, the parameters of the desired state of motion, and the motion constraints, depending on the machine kinematics and dynamics, whereas the dynamics are only represented by constant kinematic motion constraints in this book (cf. Fig. 2.4). Embedding dynamics while keeping the possibility of instantaneous reactions to unforeseen events alive is considered as a future work, which is discussed in Chap. 9.5 (p. 164). All the algorithm of Fig. 2.4 has to calculate is the respective motion set-points for the time instant T_{i+1}, which is subsequently used as a command variable for lower-level control.

Chapter 3
Mathematical Conventions and Problem Formulation

For a clear presentation of the required mathematics, it is necessary to introduce a certain nomenclature and to arrange some conventions that hold for the context of this monograph. Besides the introduction of these, we introduce a basic classification of different types and variants of on-line trajectory generation. Based on these first two sections in this chapter, the problem formulation in the introductory chapter is subsequently repeated in a formal way.

3.1 Notation and Nomenclature

This section introduces the nomenclature used in this paper, defines some important terms, and gives a first impression of how the regarded trajectories are mathematically represented.

We consider PC- or micro-controller-based systems for robot motion control, such that we assume a time-discrete overall system with a set of time instants

$$\begin{aligned} \mathcal{T} &= \{T_0, \ldots, T_i, \ldots, T_N\} \\ &\quad \text{with } \; T_i = T_{i-1} + T^{cycle} \; \text{ and } \; i \in \{1,\ldots,N\} \ , \end{aligned} \tag{3.1}$$

where T^{cycle} represents the cycle time of the system. Time-discrete values are represented by capital letters, time-continuous values by lower case letters. The position of the robotic system at instant T_i is

$$\vec{P}_i = \left({}_1P_i, \ldots, {}_kP_i, \ldots, {}_KP_i\right)^T \ , \tag{3.2}$$

where K is the number of DOFs. Velocities, accelerations, and jerks are analogously represented by $\vec{V}_i$, $\vec{A}_i$, and $\vec{J}_i$. A complete state of motion at instant T_i is described by the matrix

$$\begin{aligned} \mathbf{M}_i &= \left(\vec{P}_i, \vec{V}_i, \vec{A}_i, \vec{J}_i\right) \\ &= \left({}_1\vec{M}_i, \ldots, {}_k\vec{M}_i, \ldots, {}_K\vec{M}_i\right)^T \ , \end{aligned} \tag{3.3}$$

T. Kröger: On-Line Trajectory Generation in Robotic Systems, STAR 58, pp. 33–43.
springerlink.com

where the vector ${}_k\vec{M}_i$ is the k-th row of the matrix $\mathbf{M}_i$. Boundary conditions for motion properties are denoted as

$$\begin{aligned}\mathbf{B}_i &= \left(\vec{V}_i^{max}, \vec{A}_i^{max}, \vec{J}_i^{max}, \vec{D}_i^{max}\right)\\ &= \left({}_1\vec{B}_i, \ldots, {}_k\vec{B}_i, \ldots, {}_K\vec{B}_i\right)^T ,\end{aligned} \tag{3.4}$$

where $\vec{D}_i^{max}$ is the maximum value for the derivative of jerk at time instant T_i. A target state of motion is denoted by

$$\begin{aligned}\mathbf{M}_i^{trgt} &= \left(\vec{P}_i^{trgt}, \vec{V}_i^{trgt}, \vec{A}_i^{trgt}, \vec{J}_i^{trgt}\right)\\ &= \left({}_1\vec{M}_i^{trgt}, \ldots, {}_k\vec{M}_i^{trgt}, \ldots, {}_K\vec{M}_i^{trgt}\right)^T .\end{aligned} \tag{3.5}$$

T_N is the time instant at which $\mathbf{M}_i^{trgt}$ will be reached. As we will explain later, time-continuous polynomials are needed for the internal representation of trajectories:

$${}_k^l p_i(t) = a_4(t-\Delta t)^4 + a_3(t-\Delta t)^3 + a_2(t-\Delta t)^2 + a_1(t-\Delta t) + a_0 \tag{3.6}$$

constitutes a fourth-order polynomial describing the position progression for DOF k, which is time-shifted by Δt, and which was calculated at time instant T_i. The index l is described later in this section. Polynomials for velocity, acceleration, and jerk progressions are analogously denoted by ${}^l\vec{v}_i(t)$, ${}^l\vec{a}_i(t)$, and ${}^l\vec{j}_i(t)$. In summary, we obtain a matrix of polynomials:

$$\begin{aligned}{}^l\mathbf{m}_i(t) &= \left({}^l\vec{p}_i(t), {}^l\vec{v}_i(t), {}^l\vec{a}_i(t), {}^l\vec{j}_i(t)\right)\\ &= \left({}_1^l\vec{m}_i(t), \ldots, {}_k^l\vec{m}_i(t), \ldots, {}_K^l\vec{m}_i(t)\right)^T\end{aligned} \tag{3.7}$$

where a single row, that is, the polynomials of one single DOF k, is denoted by

$${}_k^l\vec{m}_i(t) = \left({}_k^l p_i(t), {}_k^l v_i(t), {}_k^l a_i(t), {}_k^l j_i(t)\right) . \tag{3.8}$$

Each set of motion polynomials ${}^l\mathbf{m}_i(t)$ for all K DOFs is accompanied by a set of corresponding time intervals

$${}^l\mathcal{V}_i = \left\{{}_1^l\vartheta_i, \ldots, {}_k^l\vartheta_i, \ldots, {}_K^l\vartheta_i\right\}, \text{ where } {}_k^l\vartheta_i = \left[{}_k^l t_i, {}^{(l+1)}_{\quad k} t_i\right] , \tag{3.9}$$

in which a set of polynomials ${}_k^l\vec{m}_i(t)$ for one single DOF k is valid. A whole motion trajectory $\mathcal{M}_i(t)$ is finally composed of a set of motion polynomials with according time intervals:

$$\mathcal{M}_i(t) = \left\{ \left({}^1\mathbf{m}_i(t), {}^1\mathcal{V}_i\right), \ldots, \left({}^l\mathbf{m}_i(t), {}^l\mathcal{V}_i\right), \ldots, \left({}^L\mathbf{m}_i(t), {}^L\mathcal{V}_i\right) \right\} . \tag{3.10}$$

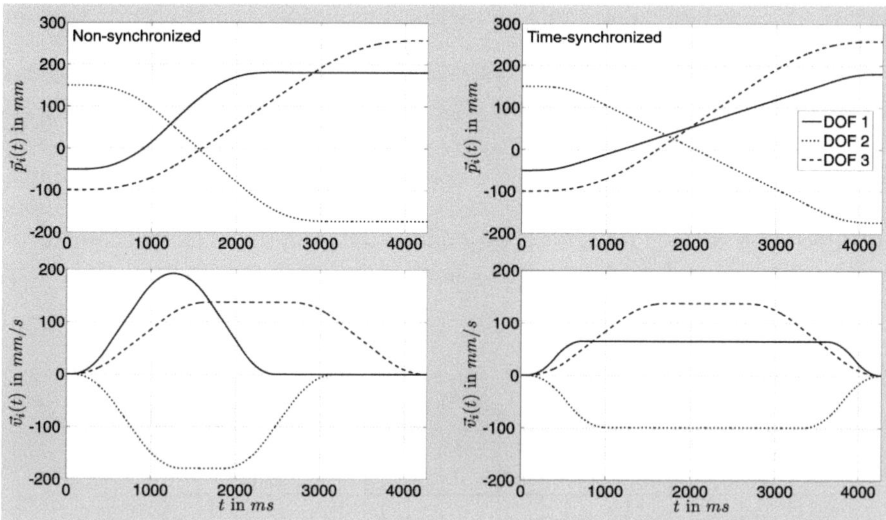

Fig. 3.1 Example of time-synchronization of a simple third-order trajectory for three translational DOFs.

Depending on the type of OTG algorithm (cf. Sec. 3.2), the initial state of motion $\mathbf{M}_0$, and the target state of motion $\mathbf{M}_0^{trgt}$, the value L determines the required number of polynomials (i.e., the number of single trajectory segments) to describe the complete trajectory from $\mathbf{M}_0$ to $\mathbf{M}_0^{trgt}$. The trajectory segments are connected to each other as:

$\forall\, k \in \{1,\ldots,K\}$:

$$\begin{aligned} {}_k^1t_i &= T_i \\ {}_k^1\vec{m}_i(T_i) &= {}_k\vec{M}_i \\ {}_k^l\vec{m}_i({}_k^lt_i) &= {}_k^{(l-1)}\vec{m}_i({}_k^lt_i) \quad \text{with } l \in \{2,\ldots,L\} \\ {}_k^L\vec{m}_i(t_i^{sync}) &= {}_k\vec{M}_i^{trgt} \ . \end{aligned} \tag{3.11}$$

From eqns. (3.9) – (3.11) we also infer

$$ {}_k^{(L+1)}t_i \equiv t_i^{sync} \ . \tag{3.12}$$

One important property of OTG is that all DOFs that are selected for trajectory-following control, have to reach their target state of motion $\mathbf{M}_i^{trgt}$ at the same time instant, namely at t_i^{sync} in order to achieve *time-synchronization*. As a consequence of this requirement, we can already state that

$$T_N - t_i^{sync} \le T^{cycle} \ , \tag{3.13}$$

that is, the desired target state of motion has been reached after a time of $N\,T^{cycle}$.

Fig. 3.1 clarifies the meaning of time-synchronization. Just to give an impression of time dimensions, T^{cycle} is in the range of one millisecond or less; the whole algorithm is designed to be applicable on a very low control level.

Now, we know how to describe a complete motion trajectory $\mathcal{M}_i(t)$ at a time instant T_i. It constitutes the minimum set of parameters to handle sets of trajectories for OTG. For a better understanding and to clarify these introductory equations, Fig. 3.2 depicts a simple (translational) motion trajectory for one single DOF k.

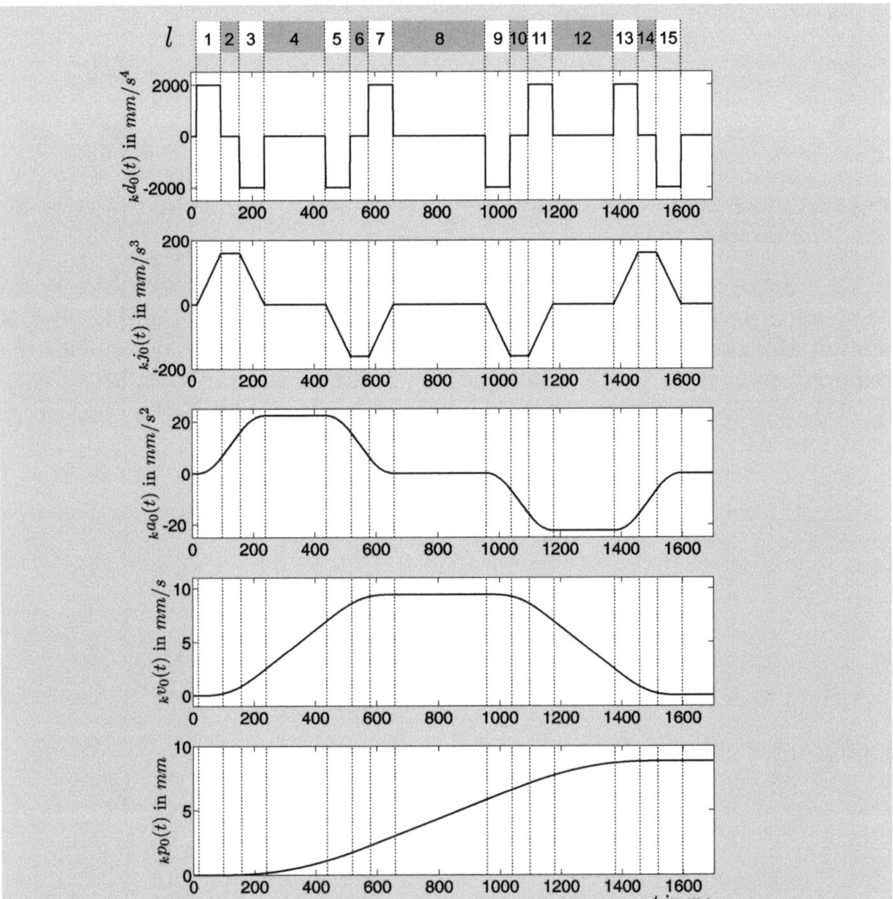

Fig. 3.2 Example of a very simple case of a fourth-order motion trajectory for one DOF k consisting of $L = 15$ matrices of polynomials ${}^l\mathbf{m}_i(t)$ (cf. eqn. 3.7). The input parameters are ${}_k\vec{M}_0 = \vec{0}$, ${}_k\vec{B}_0 = \left(9.4mm/s, 22.4mm/s^2, 160mm/s^3, 2000mm/s^4\right)$, and ${}_k\vec{M}_0^{trgt} = \left(8.8mm, 0mm/s, 0mm/s^2, 0mm/s^3\right)$ (cf. [155]).

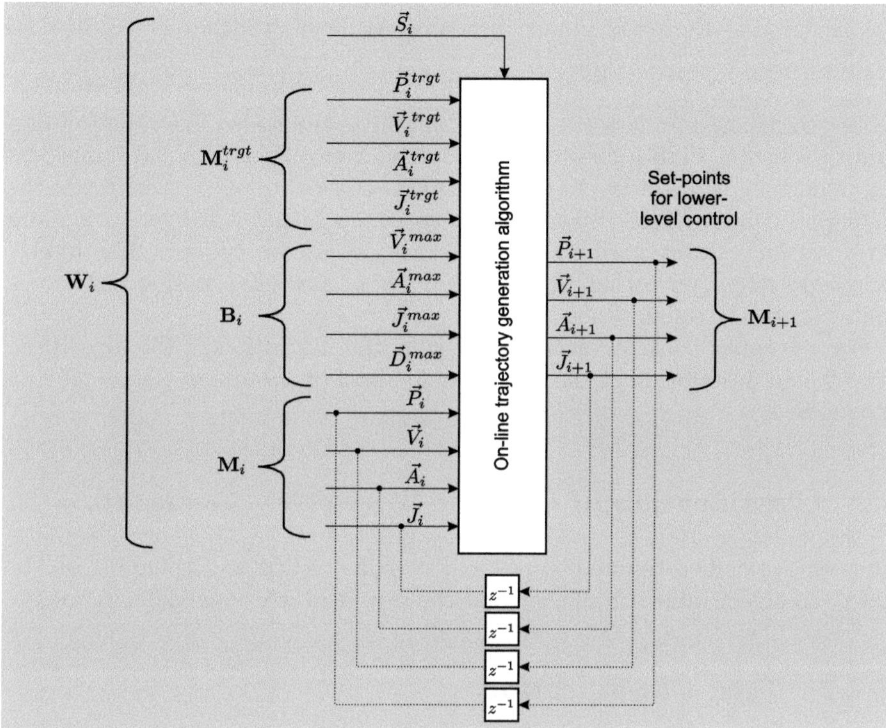

Fig. 3.3 Input and output values of the OTG algorithm for multiple DOFs (z^{-1} represents a hold element). The dotted lines indicate suggestively how the output values of the OTG are usually fed back.

For the last part of this section, we explain the accordance of these introductory descriptions with the input and output values of the OTG algorithm as it will be described in Chaps. 4 – 6. Fig. 3.3 details the input parameters

$$\begin{aligned}\mathbf{W}_i &= \left(\mathbf{M}_i, \mathbf{M}_i^{trgt}, \mathbf{B}_i, \vec{S}_i\right) \\ &= \left({}_1\vec{W}_i, \ldots, {}_k\vec{W}_i, \ldots, {}_K\vec{W}_i\right)^T \end{aligned} \tag{3.14}$$

and all output values $\mathbf{M}_{i+1}$ of the OTG algorithm, where $\mathbf{W}_i$ is a $(K \times 13)$ matrix. The selection vector

$$\vec{S}_i = \left({}_1s_i, \ldots, {}_ks_i, \ldots, {}_Ks_i\right) \quad \text{with } \vec{S}_i \in \mathbb{B}^K \tag{3.15}$$

is a Boolean vector and determines which of the K DOFs are to be controlled by the OTG. The set

$$\mathbb{B} = \{0, 1\} \tag{3.16}$$

is considered as the set of binary numbers. All DOFs that are not controlled by the OTG open-loop controller, are not considered by the algorithm.

As will be discussed in Chap. 7 (p. 105), the way the output values $\mathbf{M}_{i+1}$ are processed depends on how the OTG algorithm is embedded in a robot motion control scheme. Hence, the term *lower-level control* (cf. Fig. 3.3) considers the underlying control layers from the OTG algorithm on. For example, the algorithm can be directly used in joint space, such that it delivers command variables for a joint position controller, or it can be be part of a hybrid switched-system, such that $\mathbf{M}_{i+1}$ contributes to a desired motion state for a controller in Cartesian space.

For a complete list of variables used for the derivation of the algorithm, as well as a description of the super- and subscript meanings, please refer to the preface.

3.2 Classification of On-Line Trajectory Generators

This section defines different types and variants of OTGs. Depending on the type, the algorithmic complexity and the practical relevance differ strongly.

3.2.1 Type Classification

In general, we can denote, that the output values of the OTG algorithm are a function f of the input parameters $\mathbf{W}_i$, and that the OTG algorithm itself is *memoryless*:

$$\mathbf{M}_{i+1} = f\left(\mathbf{W}_i\right), \text{ where } f:\ \mathbb{R}^{\alpha K} \times \mathbb{B}^K \longrightarrow \mathbb{R}^{\beta K}\ . \tag{3.17}$$

α and β are type-dependent integer values, and Table 3.1 shows a summary of types of OTG. The OTG block of Fig. 3.3 — if fully connected to all input parameters and output values — corresponds to Type IX. Depending on the type, not all input and output variables are used. Type VIII, for example, does not offer to specify $\vec{J}_i^{trgt}$. For example, Fig. 3.2 shows a Type VI trajectory and Fig. 3.1 a Type III trajectory. The case $\beta = 4$ means that the position progression is described by polynomials up to the fourth order; thus, even the derivative of the jerk is limited ($\vec{D}_i^{max} \in \mathbb{R}^K$). All further types are defined analogously. To finalize Table 3.1, one may denote the trivial and practically irrelevant case with rectangular velocity profiles as Type 0 (with $\alpha = 3, \beta = 1$). Of course, it would also be possible to extend Table 3.1 by higher-order trajectories (Type X, XI, etc.), but the complexity strongly increases with increasing type numbers, such that the development of these algorithms is hardly possible, as will be figured out in Chap. 9.12 (p. 175).

A very specific version of Type I was already suggested in [146]. It works like Type II with unlimited jerks. These two types may offer sensor integration possibilities for experimental purposes, for example, in research institutions.

Table 3.1 Different types of OTG algorithms. The type number defines the set of available input parameters $\mathbf{W}_i$ and output parameters $\mathbf{M}_{i+1}$ (cf. eqns. (3.3), (3.4), (3.5), (3.14), and (3.17)).

$\forall\ (i,k) \in \mathbb{Z}\times\{1,\ldots,K\}$	${}_kV_i^{trgt}=0 \wedge {}_kA_i^{trgt}=0 \wedge {}_kJ_i^{trgt}=0$	${}_kV_i^{trgt}\in\mathbb{R} \wedge {}_kA_i^{trgt}=0 \wedge {}_kJ_i^{trgt}=0$	${}_kV_i^{trgt}\in\mathbb{R} \wedge {}_kA_i^{trgt}\in\mathbb{R} \wedge {}_kJ_i^{trgt}=0$	${}_kV_i^{trgt}\in\mathbb{R} \wedge {}_kA_i^{trgt}\in\mathbb{R} \wedge {}_kJ_i^{trgt}\in\mathbb{R}$
${}_kA_i^{max}\in\mathbb{R} \wedge {}_kJ_i^{max}=\infty \wedge {}_kD_i^{max}=\infty$	**Type I** $\alpha=5,\ \beta=2$	**Type II** $\alpha=6,\ \beta=2$	—	—
${}_kA_i^{max}\in\mathbb{R} \wedge {}_kJ_i^{max}\in\mathbb{R} \wedge {}_kD_i^{max}=\infty$	**Type III** $\alpha=7,\ \beta=3$	**Type IV** $\alpha=8,\ \beta=3$	**Type V** $\alpha=9,\ \beta=3$	—
${}_kA_i^{max}\in\mathbb{R} \wedge {}_kJ_i^{max}\in\mathbb{R} \wedge {}_kD_i^{max}\in\mathbb{R}$	**Type VI** $\alpha=9,\ \beta=4$	**Type VII** $\alpha=10,\ \beta=4$	**Type VIII** $\alpha=11,\ \beta=4$	**Type IX** $\alpha=12,\ \beta=4$

But due to the non-smooth bang-bang or trapezoidal trajectory characteristics [126], they are not relevant for industrial applications, as has already been stated in [120]. To achieve long lifetimes of mechanical systems, a jerk limitation is required (Types III to V). Furthermore, smooth trajectories whose jerk or even the derivative of jerk is limited lead to better trajectory-following behavior and to less excitation of structural natural frequencies in the system. Compared to Type III, Type IV and V additionally provide the possibility of specifying target velocity vectors and/or target acceleration vectors in space, which is important for the consideration of system dynamics. Target acceleration vectors are considered as the curvature of target velocity vectors and become relevant for planned switchings between control spaces, which will be discussed in Chap. 9.12 (p. 175).

3.2.2 Variant Classification

The distinction between different variants, Variant A and Variant B, is basically not required at all, but this additional classification simplifies explanations of the algorithmic details of the OTG algorithms presented in the next three chapters.

Regarding different variants of OTG, we distinguish between constant and non-constant boundary values and define two variants, A and B:

- *Variant A*: $\mathbf{B}_i = \mathbf{const} \ \forall \ i \in \mathbb{Z}$
- *Variant B*: $\mathbf{B}_i \neq \mathbf{const}$.

The second slightly more advanced variant is the one that will be relevant for practical implementations. Variant A only helps to explain the algorithm in a more comprehensive way.

3.3 Formal Problem Formulation

Based on the previous two sections, we now render the problem formulation of Chap. 1.1.3 more precisely. Therefore, we start with the definition of the term *time-optimality* and distinguish between two different kinds:

Definition 3.1 (Kinetic time-optimality). *A system is transferred from an initial state of motion $\mathbf{M}_i$ at instant T_i into a desired target state of motion $\mathbf{M}_i^{trgt}$ within the shortest possible time without any consideration of couplings between individual DOFs.*

If we consider a system with K DOFs, that is, K actuators, each actuator possess a maximum force (or torque). Depending on the mass (or the moment of inertia), a constant maximum linear (or angular) acceleration value results, if Newton's Second Law of Motion is applied. This acceleration value can be calculated for each of the K DOFs, and all acceleration values together constitute $\vec{A}_i^{max}$, that is one of the input vectors of the OTG algorithm.

Definition 3.2 (Dynamic time-optimality). *A system is transferred from an initial state of motion $\mathbf{M}_i$ at instant T_i into a desired target state of motion $\mathbf{M}_i^{trgt}$ within the shortest possible time with consideration of the whole system dynamics.*

For the OTG algorithm itself, that is, for all calculations within the block of Fig. 3.3, only *kinetic time-optimality* is regarded. The important consequence of kinetic time-optimality is, that all K DOFs can be considered as *linearly independent*, that is, *decoupled*. How robot kinematics and *dynamics* can be combined with this approach is described in Chap. 9.5 (p. 164).

For a clear and unambiguous specification of the problem to be solved by an on-line trajectory generation algorithm, we distinguish between four different criteria (i) – (iv), which all have to be met by the algorithm.

(i) Criterion for Time-Optimality

The problem to be solved in this work is to calculate all parameters of $\mathcal{M}_i(t)$ at instant T_i, such that all selected DOFs reach their target state of motion $\mathbf{M}_i^{trgt}$ in the shortest possible time t_i^{sync}:

$\forall\ (k,l)\ \in\ \{1,\ldots,K\}\times\{1,\ldots,L\}:$

$$\left.\begin{array}{l}\left|{}_k^l v_i(t)\right|\ \leq\ {}_kV_i^{max}\\ \left|{}_k^l a_i(t)\right|\ \leq\ {}_kA_i^{max}\\ \left|{}_k^l j_i(t)\right|\ \leq\ {}_kJ_i^{max}\\ \left|{}_k^l d_i(t)\right|\ \leq\ {}_kD_i^{max}\end{array}\right\}\ \text{with}\ t\ \in\ \left[{}_k^l t_i,\ {}_k^{(l+1)}t_i\right]\ , \tag{3.18}$$

such that

$$t_i^{sync}\ \longrightarrow\ \min\ . \tag{3.19}$$

t_i^{sync} is a parameter of $\mathcal{M}_i(t)$ (cf. eqns.(3.9) – (3.12)).

(ii) Criterion for Time-Synchronization

The time t_i^{sync} plays a very central role in this work. To ensure that $\vec{V}_i^{trgt}$, $\vec{A}_i^{trgt}$, and $\vec{J}_i^{trgt}$ are reached in $\vec{P}_i^{trgt}$ at t_i^{sync}, we can repeat the aspect of time-synchronization (eqn. (3.11)) again:

$$_k^L\vec{m}_i(t_i^{sync})\ =\ {}_k\vec{M}_i^{trgt}\quad \forall\ k\ \in\ \{1,\ldots,K\}\ . \tag{3.20}$$

(iii) Criterion for Motion Constraints

Equation (3.18) guarantees that the motion variables of $\mathcal{M}_i(t)$ are kept within their respective bounds of $\mathbf{B}_i$. In Variant A ($\mathbf{B}_i = \mathbf{const}\ \forall\ i\ \in\ \mathbb{Z}$), this is clear and simple, but in Variant B, it of course can happen that:

- one or more of the elements of $\mathbf{M}_i$ exceed their bounds of $\mathbf{B}_i$ (e.g., $_1V_i = 5\,m/s$ and $_1V_i^{max} = 4\,m/s$), or
- one or more of the elements of $\mathbf{M}_i$ are within their bounds, but a future excess of their bounds $\mathbf{B}_i$ is unavoidable (e.g., $_2V_i = 4\,m/s$, $_2V_i^{max} = 5\,m/s$, $_2A_i = 2\,m/s^2$, and $_2J_i^{max} = 1\,m/s^3$: Even if $-_2J_i^{max}$ would be applied permanently, the velocity $_2v(t)$ would unavoidably increase to $6\,m/s$).

If one of these two conditions is true at T_i, the OTG Variant B algorithm has to determine and calculate further trajectory segments, which are applied upstream to the ones that are calculated to satisfy eqns. (3.18) and (3.19). These Variant B trajectory segments have to lead the variables of $\mathbf{M}_i$ that have exceeded their respective boundary values of $\mathbf{B}_i$, or that will exceed their respective boundary values of $\mathbf{B}_i$ back into their limits within the shortest possible time, such that these values can always be kept within their bounds.

We take arbitrary values for $\mathbf{W}_i$ for granted; the only (trivial) constraint for $\mathbf{W}_i$ is given by

$\forall\, k \in \{1,\ldots,K\}:$

$$_kV_i^{trgt} \leq {_kV_i^{max}} \wedge {_kA_i^{trgt}} \leq {_kA_i^{max}} \wedge {_kJ_i^{trgt}} \leq {_kJ_i^{max}}\,. \qquad (3.21)$$

As already stated in eqn. (3.19), the challenge is to find a motion trajectory $\mathcal{M}_i(t)$ that transfers the state of motion from $\mathbf{M}_i$ to $\mathbf{M}_i^{trgt}$ within the shortest possible time t_i^{sync}. But for the current control cycle at T_i, only $\mathbf{M}_{i+1}$ is needed, because in the next control cycle, we might have completely new input values $\mathbf{W}_{i+1}$ due to an unforeseen event or switching action. Hence, only $\mathbf{M}_{i+1}$ is forwarded to the output. These values are then used as input values for lower-level control.

(iv) Criterion for Consistency

This leads to the interesting requirement, that if the input values $\mathbf{M}_i^{trgt}$, $\mathbf{B}_i$, and $\vec{S}_i$ remain constant for $i \in \{0,\ldots,N\}$ (cf. eqn. (3.13)), and if the output values $\mathbf{M}_{i+1}$ are directly fed back as input values for the following control cycle (cf. dotted lines of Fig. 3.3), then the following consistency criterion must be fulfilled.

If $\mathbf{M}_i^{trgt}$, $\mathbf{B}_i$, and $\vec{S}_i$ remain constant for $i \in \{0,\ldots,N\}$, then:

- the value of the synchronization time must remain constant during the whole trajectory execution, that is, $t_i^{sync} = \text{const}\ \forall\, i \in \{0,\ldots,N\}$. This fact is relevant for time-synchronization, such that $\vec{V}_i^{trgt}$, $\vec{A}_i^{trgt}$, and $\vec{J}_i^{trgt}$ are coinstantaneously reached in $\vec{P}_i^{trgt}$ at t_i^{sync};
- all trajectories $\mathcal{M}_u(t)$ with $u \in \{1,\ldots,N\}$ must exactly fit into the one of $\mathcal{M}_0(t)$; and
- furthermore, any trajectory $\mathcal{M}_u(t)$ with $u \in \{1,\ldots,N\}$ must exactly fit into all previously calculated motion trajectories $\mathcal{M}_v(t)$ with $v \in \{0,\ldots,u-1\}$.

This criterion guarantees that the resulting trajectories of the OTG algorithm behave exactly as off-line planned ones, for example, bang-bang and/or trapezoidal acceleration profiles (cf. [29, 43, 126, 196]).

3.4 Summary

To summarize this chapter briefly: We introduced a dedicated notation and described the task of OTG very generally without offering solutions. The problem was formulated formally, whereas four different criteria (i) – (iv) were specified. Each single criterion is essential and has to be met by the OTG algorithm. The final and desired algorithm consists of three steps, and a part of the first step is also applicable to and relevant for one-DOF systems. In order to derive the OTG algorithm step by step, we first solve a part of the addressed problem in one-dimensional space in Chap. 4. Chap. 5 is based on Chap. 4 and subsequently gives the complete solution in multi-dimensional space. The Type IV OTG algorithm includes the Types I – III (cf. Table 3.1) and is the highest type that has been developed by the author. This type will be used to produce, illustrate, and explain examples within this monograph. Each chapter explains the concept of OTG in a very generic way and illustrates the ideas by means of Type IV.

Chapter 4
Solution for One Degree of Freedom

This work introduces both, an OTG algorithm for one-DOF systems and an OTG algorithm for systems with multiple DOFs, as they are common in the field of robotics. Also the one-DOF solution that is presented in this chapter delivers new significant advantages for servo drive control and the technology of frequency inverters. Furthermore, it simplifies the comprehension of this work, since we introduce the OTG algorithm step by step.

4.1 Generic Algorithm for On-Line Trajectory Generation

4.1.1 Problem Formulation for One-DOF Systems

The problem formulation for one single DOF is very similar to the one for multiple DOFs (cf. Chap. 3.3, p. 40). The input values of the multi-DOF OTG algorithm, which are vectors, become scalars, and the matrices $\mathbf{W}_i$, $\mathbf{M}_i^{trgt}$, $\mathbf{B}_i$, $\mathbf{M}_i$, $\mathbf{M}_{i+1}$ become vectors (rows of former matrices), as depicted in Fig. 4.1.

We take arbitrary values for the elements of

$$\vec{M}_i = (P_i, V_i, A_i, J_i) \tag{4.1}$$

$$\vec{M}_i^{trgt} = \left(P_i^{trgt}, V_i^{trgt}, A_i^{trgt}, J_i^{trgt}\right) \tag{4.2}$$

$$\vec{B}_i = \left(V_i^{max}, A_i^{max}, J_i^{max}, D_i^{max}\right) \tag{4.3}$$

$$\vec{W}_i = \left(\vec{M}_i, \vec{M}_i^{trgt}, \vec{B}_i, S_i\right) \tag{4.4}$$

for granted. If the value of S_i is '1'[1] at a time instant T_i, it is our aim to calculate a trajectory $\mathcal{M}_i$ that transfers the system from its current state of

[1] For the rest of this chapter, we assume $S_i = 1$.

T. Kröger: On-Line Trajectory Generation in Robotic Systems, STAR 58, pp. 45–68.
springerlink.com

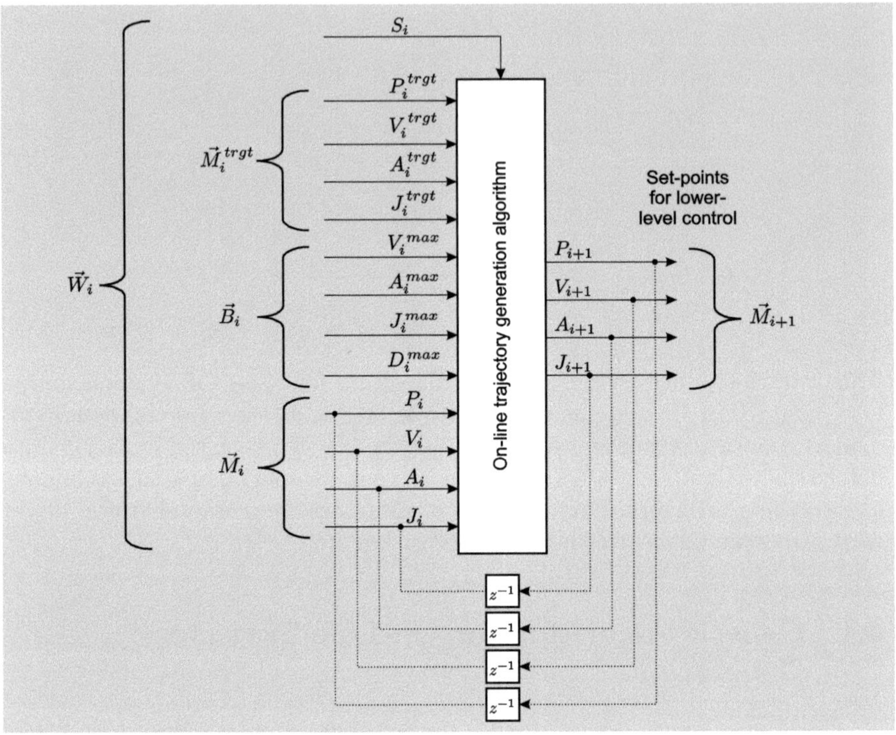

Fig. 4.1 Input and output values of the Type IX OTG algorithm for one DOF. (cf. Fig. 3.3, p. 37).

motion $\vec{M}_i$ into its current target state of motion $\vec{M}_i^{trgt}$ under consideration of the boundary constraints $\vec{B}_i$ and in the minimum possible time t_i^{min}. In accordance with eqn. (3.10), a trajectory $\mathcal{M}_i$ for one DOF consists of L trajectory segments

$$ {}^{l}\vec{m}_i(t) = \left({}^{l}p_i(t), {}^{l}v_i(t), {}^{l}a_i(t), {}^{l}j_i(t)\right) \quad \text{with } l \in \{\Lambda + 1, \ldots, L\} \qquad (4.5) $$

and corresponding time intervals

$$ {}^{l}\vartheta_i = \left[{}^{l-1}t_i, {}^{l}t_i\right] \quad \text{with } l \in \{\Lambda + 1, \ldots, L\} \,. \qquad (4.6) $$

If Variant A is applied (i.e., $\vec{B}_i$ is constant for all $i \in \mathbb{Z}$),

$$ \Lambda = 0 \qquad (4.7) $$

holds. Λ specifies the number of intermediate trajectory segments that are selected by the B-Variants; this issue is described in the next subsection. If we know all parameters of $\mathcal{M}_i$, we can simply calculate the desired output values for lower-level control

$$\vec{M}_{i+1} = (P_{i+1}, V_{i+1}, A_{i+1}, J_{i+1}) \ . \tag{4.8}$$

The calculation of eqn. (4.8) is described in a general way in the next subsection and will be concretized by means of the Type IV OTG algorithm in Sec. 4.2.

4.1.2 Generic Solution

At this point we distinguish between Variant A, in which the values of $\vec{B}_i$ are considered as constant, and Variant B, in which the elements of $\vec{B}_i$ may vary over time.

Generic Solution: Variant A

The key idea of this work is that there is a *finite* set of possible motion profiles, of which one transfers one single selected DOF from the initial state of motion $\vec{M}_i$ to its target state of motion $\vec{M}_i^{trgt}$ within the shortest possible time t_i^{min} (time-optimally). This finite set is denoted by[2]

$$\mathcal{P}_{Step1} = \left\{ {}^{1}\Psi^{Step1}, \ldots, {}^{r}\Psi^{Step1}, \ldots, {}^{R}\Psi^{Step1} \right\} \ , \tag{4.9}$$

where R is the number of elements in $\mathcal{P}_{Step1}$ and depends on the type of OTG. A concrete profile is denoted by ${}^{r}\Psi^{Step1}$. The kind of motion profile considered depends on the type of OTG used (cf. Table 3.1, p. 39):

$$\begin{aligned} &\textit{Types I-II}: \text{Velocity profiles} && (\beta = 2) \\ &\textit{Types III-V}: \text{Acceleration profiles} && (\beta = 3) \\ &\textit{Types VI-IX}: \text{Jerk profiles} && (\beta = 4) \ . \end{aligned}$$

The first task for the calculation of $\mathcal{M}_i$ is the execution of a function f

$$f: \ \mathbb{R}^{\alpha} \longrightarrow \mathcal{P}_{Step1} \ , \tag{4.10}$$

which is represented by a decision tree in order to select the motion profile that leads to *the* time-optimal trajectory, and that simultaneously enables us to determine t_i^{min} for the considered DOF. Once *the* time-optimal profile Ψ_i^{Step1} has been selected, a system of nonlinear equations can be set up and solved to calculate the coefficients of all polynomials ${}^{l}\vec{m}_i(t) \ \forall \ l \ \in \{1, \ldots, L\}$ and the respective time intervals ${}^{l}\vartheta_i \ \forall \ l \ \in \ \{1, \ldots, L\}$. As a result,

[2] The algorithm for the multi-dimensional case consists of three steps. As the set for the one-dimensional case is the same as the one for the first step in the multi-dimensional case, we use the same notation in order to keep the representation unique.

$$ {}^{L}t_i = t_i^{min} \tag{4.11} $$

holds. Each motion profile ${}^{r}\Psi^{Step1}$ with $r \in \{1, \ldots, R\}$ leads to a system of nonlinear equations, and each system is solvable for a certain domain ${}^{r}\mathcal{D}_{Step1}$ with

$$ {}^{r}\mathcal{D}_{Step1} \subset \mathbb{R}^{\alpha} \ . \tag{4.12} $$

For the decision tree that implements eqn. (4.10) it is *absolutely essential* that

$$ \bigcup_{r=1}^{R} {}^{r}\mathcal{D}_{Step1} \ \equiv \ \mathbb{R}^{\alpha} \tag{4.13} $$

holds. If this is not the case, the tree is erroneous, and the algorithm will not work for certain input parameters, which would be unacceptable for its practical application. In the work of Broquère, Sidobre, and Herrera-Aguilar [38] as well as in the contribution of Liu [167], which suggest similar approaches for one-DOF systems, eqn. (4.13) does not hold, and, hence, these approaches are not applicable in general, but only for some (practically hardly relevant) special cases.

As an alternative to decision trees for the selection of the correct motion profile Ψ_i^{Step1}, one could set up R systems of nonlinear equations, calculate all solutions, take all valid solutions for t_i^{min}, and choose the minimum one. But this procedure is computationally very expensive, too expensive, in fact, especially if low values for T^{cycle} are desired.

After we have calculated all L trajectory segments to describe $\mathcal{M}_i$, we only have to find the valid time interval ${}^{\hat{l}}\vartheta_i$ with $\hat{l} \in \{1, \ldots, L\}$, such that

$$ {}^{\hat{l}}t_i \leq \left(T_i + T^{cycle}\right) \leq {}^{\hat{l}+1}t_i \tag{4.14} $$

is satisfied (cf. eqns. (3.9) and (3.10), p. 34). The output values $\vec{M}_{i+1}$ can finally be calculated by

$$ \vec{M}_{i+1} = {}^{\hat{l}}\vec{m}_i \left(T_i + T^{cycle}\right) . \tag{4.15} $$

Up to now, we only have considered Variant A, which requires constant values of $\vec{B}_i$. The proposed procedure will now be extended by Variant B to make this approach generally valid.

General Solution: Variant *B*

Variant B works in the same way as A does, but here a further decision tree is connected upstream of the Variant A decision tree. If the boundary values $\vec{B}_i$ are functions of time, the OTG Variant B has to be applied. The case in which one or more elements of the initial motion state values $\vec{M}_i$ exceed their corresponding constraints of $\vec{B}_i$ has to be considered in this variant, and it may happen at any discrete time instant T_i with $i \in \mathbb{Z}$, in which

$$\begin{aligned} &|V_i| > V_i^{max} \quad \text{and/or} \quad |A_i| > A_i^{max} \quad \text{and/or} \\ &|J_i| > J_i^{max} \quad \text{and/or} \quad |D_i| > D_i^{max} \ . \end{aligned} \tag{4.16}$$

Furthermore, motion states $\vec{M}_i$ may occur, which are within their respective bounds $\vec{B}_i$ at instant T_i, but which will lead to an unavoidable future exceeding of $\vec{B}_i$ at a time instant T_{i+u}:

$$\begin{aligned} &|V_{i+u}| > V_i^{max} \quad \text{and/or} \quad |A_{i+u}| > A_i^{max} \quad \text{and/or} \\ &|J_{i+u}| > J_i^{max} \quad \text{and/or} \quad |D_{i+u}| > D_i^{max} \quad \text{with } u \in \mathbb{N}\backslash\{0\} \ . \end{aligned} \tag{4.17}$$

Equations (4.16) and (4.17) correspond to the criteria of Chap. 3.3 (p. 41). If one or more of the cases in eqns. (4.16) and (4.17) are true, the Variant B decision tree has to select and parameterize intermediate trajectory segments ${}^l\vec{m}_i$ with $l \in \{1,\ldots,\Lambda\}$ with corresponding time intervals ${}^l\vartheta_i$ with $l \in \{1,\ldots,\Lambda\}$ (cf. eqns. (4.5) and (4.6)) in order to guide these values back into their bounds, and furthermore to bring the system into a state of motion after which the motion variables can be kept within the boundary values of $\vec{B}_i$. Therefore, this set of intermediate trajectory segments is determined and executed prior to the segments of Variant A: ${}^l\vec{m}_i$ with $l \in \{\Lambda+1,\ldots,L\}$ and ${}^l\vartheta_i$ with $l \in \{\Lambda+1,\ldots,L\}$. Corresponding to the type of OTG, velocity profiles (Type I, II), acceleration profiles (Type III – V), or jerk profiles (Type VI – IX) are applied, respectively.

4.2 Solution for Type IV

Compared to the general OTG algorithm of the previous section, this section exemplarily introduces one concrete type of OTG: Type IV. This first leads us to the following general parameters for the algorithm: $\alpha = 8$, $\beta = 3$, $A_i^{trgt} = 0 \wedge J_i^{trgt} = 0 \ \forall \ i \in \mathbb{Z}$, and D_i^{max} may contain infinite values for any $i \in \mathbb{Z}$ (cf. Table 3.1, p. 39). In simpler words, this type generates kinetically time-optimal and time-synchronized trajectories whose velocity, acceleration, and jerk values are limited, and the specification of target velocities V_i^{trgt}, which are reached in P_i^{trgt}, is possible.

Fig. 4.2 illustrates all input and output values of the Type IV OTG algorithm. In order to facilitate the understanding of this section and to enable a better understanding, App. D (p. 201) additionally explains the problem that is solved here in a non-scientific and simple way.

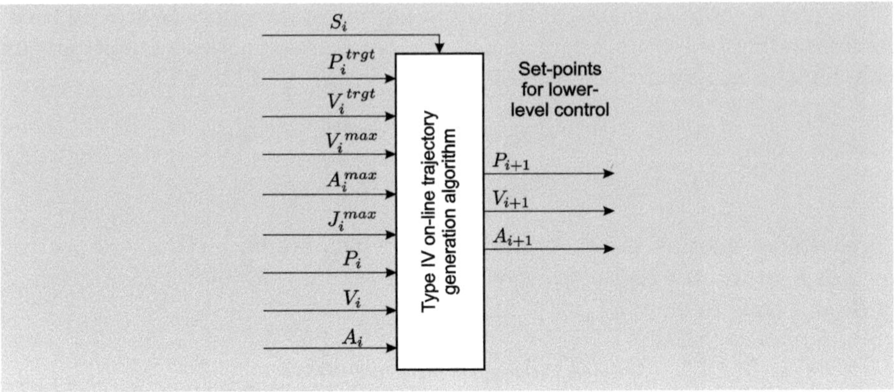

Fig. 4.2 Input and output values of the Type IV OTG algorithm.

4.2.1 Type IV, Variant A

The goal of this subsection is to calculate all trajectory parameters ${}^{l}\vec{m}_i(t)$ with $l \in \{1,\ldots,L\}$ and corresponding time intervals ${}^{l}\vartheta_i \; \forall \; l \in \{1,\ldots,L\}$, in short: $\mathcal{M}_i$. In Variant A, we assume constant boundary values $\vec{B}_i$, that is,

$$V_i^{max} = \text{const} \wedge A_i^{max} = \text{const} \wedge J_i^{max} = \text{const} \; \forall \; i \in \mathbb{Z} \; . \qquad (4.18)$$

According to the previous chapter, the Type IV OTG algorithm requires the selection of an acceleration profile, which enables us to set up a system of

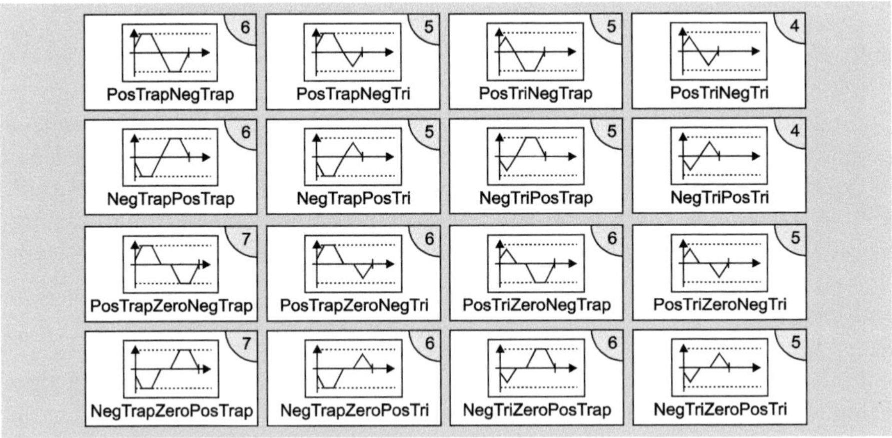

Fig. 4.3 A subset of the acceleration profile set $\mathcal{P}_{Step1}$ of Type IV, Variant A. The dotted horizontal line indicates the maximum acceleration values, and the numbers in the top right corner indicate the numbers of trajectory segments L for the respective profile.

nonlinear equations in order to calculate the parameters of $\mathcal{M}_i$. Fig. 4.3 shows a subset of the possible acceleration profiles $\mathcal{P}_{Step1}$. Compared to triangle (*Tri*) profiles, trapezoid (*Trap*) profiles always reach the maximum acceleration value A_i^{max}. All profiles that do not contain a *zero*-acceleration phase do not reach the maximum velocity V_i^{max}.

Determining Ψ_i^{Step1}

In the following, we answer the question: How can we determine *the* time-optimal acceleration profile Ψ_i^{Step1} at a time instant T_i, that is, how can we select the element of $\mathcal{P}_{Step1}$ that leads to t_i^{min}? Due to the high complexity, only a small cutout of the Type IV decision tree is shown in Fig. 4.4. The tree actually acts as function

$$f:\ \mathbb{R}^8 \longrightarrow \mathcal{P}_{Step1} \tag{4.19}$$

and determines *the* time-optimal acceleration profile Ψ_i^{Step1}. As described by eqn. (4.20), it is essential, that the tree covers the whole input domain of the Type IV, Variant *A* OTG algorithm:

$$\bigcup_{r=1}^{R} {}^{r}\mathcal{D}_{Step1} \;=\; \mathbb{R}^8 \tag{4.20}$$

where ${}^{r}\mathcal{D}_{Step1} \subset \mathbb{R}^8$ denotes the domain in which the system of equations for the acceleration profile ${}^{r}\Psi^{Step1}$ is solvable. If eqn. (4.20) does not hold, the OTG algorithm cannot be used as an open-loop controller, because there would remain motion states, for which the algorithm would not be able to calculate $\mathcal{M}_i$ and thus $\vec{M}_{i+1}$.

This paragraph discusses the decision tree of Fig. 4.4 in detail. Decision *1A.001*[3] checks, whether the current acceleration value A_i is positive or negative. The left branch is designed for positive values of A_i. Decision *1A.002* calculates the velocity value if we bring A_i down to zero (by applying J_i^{max}). If this value is less than V_i^{trgt}, the left branch is taken. Hence, we already know that $-V_i^{max} \leq V_i \leq V_i^{trgt}$. It is our aim to reach V_i^{trgt}, and Decision *1A.003* checks whether a triangle or a trapezoid profile is required for this. If a trapezoid profile is required, Decision *1A.004* checks whether the position would be less or greater than the target position P_i^{trgt}. If we the resulting position value is still less than (i.e., we are still faced with) P_i^{trgt}, we need to increase the trapezoid acceleration profile, that is, we definitely need a *PosTrap...Neg...* profile. Decision *1A.005* then checks the velocity value that would be achieved if we accelerate to $+V_i^{max}$ by applying a trapezoid profile and subsequently

[3] Part A of Step 1 of the OTG algorithm for the multi-dimensional case uses the same decision tree. We use the same notation (1*A*) in order to keep the representation consistent again.

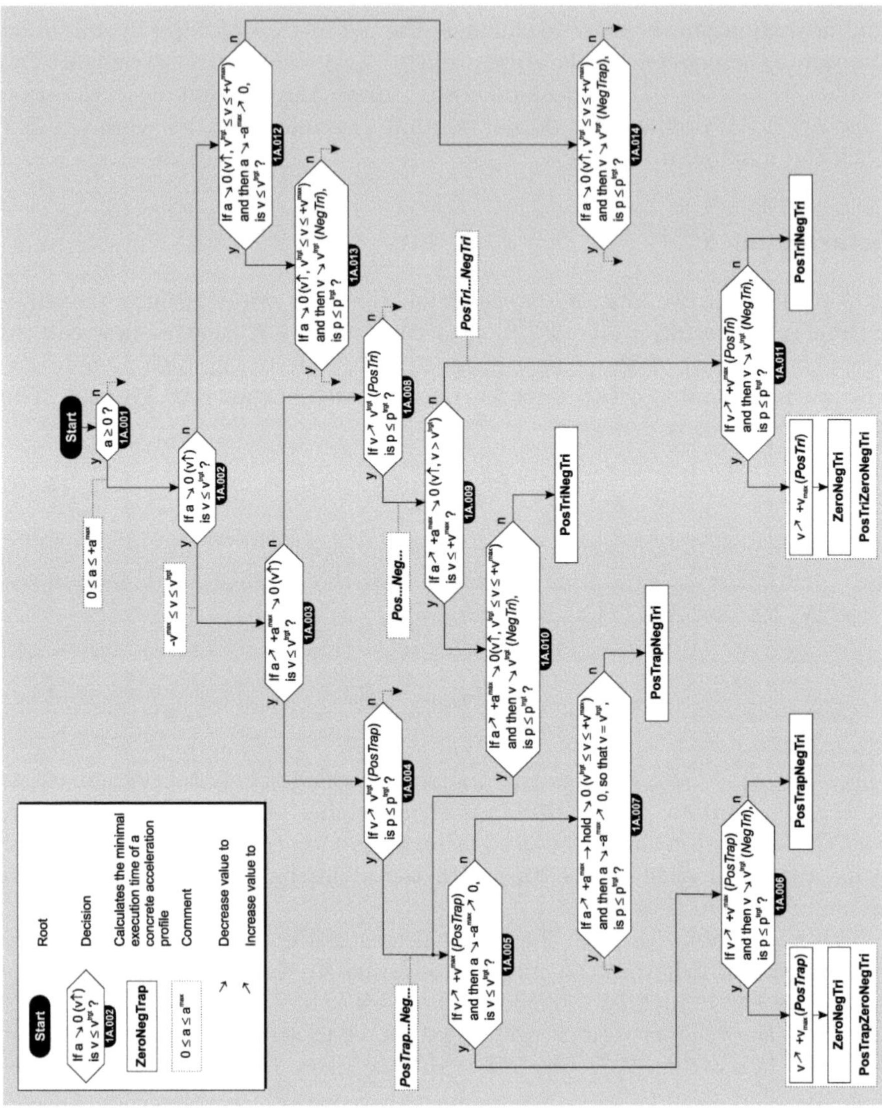

Fig. 4.4 Cutout of the Type IV, Variant *A* decision tree to determine an acceleration profile Ψ_i^{Step1}, that leads to the minimum-time solution t_i^{min}.

decelerate with a negative triangular profile that reaches $-A_i^{max}$ exactly. If the resulting velocity value is less than V_i^{trgt}, we definitely need a negative triangle profile as the second part of the composed acceleration profile. Now, we only have to check whether $+V_i^{max}$ can be reached in order to obtain V_i^{trgt} in P_i^{trgt} (Decision *1A.006*). All further decisions work analogously to the described ones. For the development of this kind of decision tree, it is absolutely

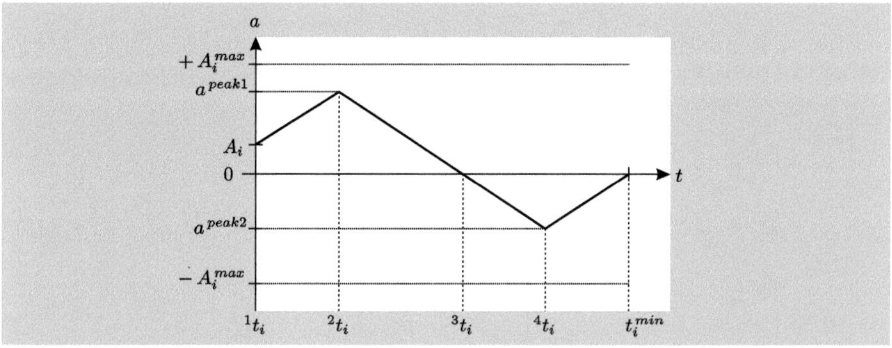

Fig. 4.5 The *PosTriNegTri* profile with all relevant variables, such that a system of equations represented by eqns. (4.21) – (4.32) can be set up and solved in order to calculate the parameters of $\mathcal{M}_i$.

essential that a tree covers the whole input domain (here: $\mathbb{R}^8$), such that the algorithm can work with arbitrary input values $\vec{W}_i$.

Parameterizing Ψ_i^{Step1} (Example: $\Psi_i^{Step1} = PosTriNegTri$)

Once we know the correct acceleration profile Ψ_i^{Step1} for the trajectory at instant T_i, we can set up a system of equations that corresponds to this profile. Depending on the profile, the procedure of parameterizing Ψ_i^{Step1} differs. In this section, we only derive the procedure for one concrete profile, that is, the *PosTriNegTri* profile (top right in Fig. 4.3). For the practical realization, all R acceleration profiles have to be considered, and respective algorithms have to be implemented, of course. Fig. 4.5 shows this *PosTriNegTri*-acceleration profile and all relevant variables, such that we can set up a system of equations to calculate the parameters of $\mathcal{M}_i$. In this simple case, we only need $L = 4$ trajectory segments. The system of equations can be directly derived from Fig. 4.5 and is given by eqns. (4.21) – (4.32). In this full-length way, we obtain 12 equations with 12 unknown variables t_i^{min}, ${}^2t_i$, ${}^3t_i$, ${}^4t_i$, ${}^2v_i$, ${}^3v_i$, ${}^4v_i$, ${}^2p_i$, ${}^3p_i$, ${}^4p_i$, a_i^{peak1}, and a_i^{peak2}:

$$ {}^2t_i - T_i = \frac{\left(a^{peak1} - A_i\right)}{J_i^{max}} \tag{4.21} $$

$$ {}^3t_i - {}^2t_i = \frac{a^{peak1}}{J_i^{max}} \tag{4.22} $$

$$ {}^4t_i - {}^3t_i = -\frac{a^{peak2}}{J_i^{max}} \tag{4.23} $$

$$ t_i^{min} - {}^4t_i = -\frac{a^{peak2}}{J_i^{max}} \tag{4.24} $$

$$ {}^{2}v_i - V_i = \frac{1}{2}\left({}^{2}t_i - T_i\right)\left(A_i + a^{peak1}\right) \tag{4.25} $$

$$ {}^{3}v_i - {}^{2}v_i = \frac{1}{2}\left({}^{3}t_i - {}^{2}t_i\right) a^{peak1} \tag{4.26} $$

$$ {}^{4}v_i - {}^{3}v_i = \frac{1}{2}\left({}^{4}t_i - {}^{3}t_i\right) a^{peak2} \tag{4.27} $$

$$ V_i^{trgt} - {}^{4}v_i = \frac{1}{2}\left(t_i^{min} - {}^{4}t_i\right) a^{peak2} \tag{4.28} $$

$$ \begin{aligned} {}^{2}p_i - P_i = {} & V_i\left({}^{2}t_i - T_i\right) + \frac{1}{2} A_i \left({}^{2}t_i - T_i\right)^2 \\ & + \frac{1}{6} J_i^{max}\left({}^{2}t_i - T_i\right)^3 \end{aligned} \tag{4.29} $$

$$ \begin{aligned} {}^{3}p_i - {}^{2}p_i = {} & {}^{2}v_i\left({}^{3}t_i - {}^{2}t_i\right) + \frac{1}{2} a^{peak1}\left({}^{3}t_i - {}^{2}t_i\right)^2 \\ & - \frac{1}{6} J_i^{max}\left({}^{3}t_i - {}^{2}t_i\right)^3 \end{aligned} \tag{4.30} $$

$$ {}^{4}p_i - {}^{3}p_i = {}^{3}v_i\left({}^{4}t_i - {}^{3}t_i\right) - \frac{1}{6} J_i^{max}\left({}^{4}t_i - {}^{3}t_i\right)^3 \tag{4.31} $$

$$ \begin{aligned} P_i^{trgt} - {}^{4}p_i = {} & {}^{4}v_i\left(t_i^{min} - {}^{4}t_i\right) + \frac{1}{2} a^{peak2}\left(t_i^{min} - {}^{4}t_i\right)^2 \\ & + \frac{1}{6} J_i^{max}\left(t_i^{min} - {}^{4}t_i\right)^3 \end{aligned} \tag{4.32} $$

Numerical Issues

Although these systems of equations are nonlinear for all elements of the set $\mathcal{P}_{Step1}$, we are able to find closed-form analytical solutions by employing computer algebra programs; this would be the straightforward way. However, these expressions become very large, not all solutions are valid, and problems with numerical stability appear. Even after appropriate simplifications and a reduction of the problem to a root-finding problem of quartic equations, it is not possible to solve this problem robustly and with a consistently high accuracy [111, 245]. In [111] robustness test results of different methods for root-finding of quartics are shown, and none of the methods work reliably. For this work, different methods (Ferrari, Neumark, and Yacoub) were implemented to find the real roots of quartics in order to solve eqns. (4.21) – (4.32), but in particular at the boundary areas of the domains ${}^{r}\mathcal{D}_{Step1}$ with $r \in \{1, \ldots, R\}$, the solutions tend to contain complex numbers due to numerical inaccuracies. Another problem is that there may be up to four real solutions for one system of equations, and it becomes difficult to extract the correct one, especially if the solutions are very close to each other. When random floating-point numbers $[-1000, 1000]$ were used for the elements of $\vec{W}_i$, a wrong solution was calculated every $\approx 60{,}000$ cycles, which is of course unacceptable in practice. Another indication that the problem of root-finding of quartics cannot be solved robustly is that neither the libraries of the

Numerical Algorithms Group (NAG) [257] nor the *GNU Scientific Library* (GSL) [100] nor the library of Press et al. [215] provide solutions for this problem.

The next idea was to perform a QR decomposition [245, 251] for the problem of quartic polynomials. With this method, it was possible to find solutions for the quartic root-finding problem, but this led to another problem: real-time capability. If a certain accuracy has to be achieved, the number of required iterations depends on the condition number of the factorization matrix [215]; and the condition number in turn depends on the input values $\vec{W}_i$, which are arbitrary. Furthermore, the problem that several solutions are found, which may lie very close to each other, remains.

As a result, the usage of analytical solution methods is not a way to success, and the following question remains unanswered: How can we find a solution method for the systems of equations of $\mathcal{P}^{Step1}$ that works robustly for any combination of input variables $\vec{W}_i$? In the following, a method is proposed that complies with this requirement, and that enables an efficient and real-time capable solution of this particular problem.

The eqns. (4.21) – (4.32) can be transformed to a one-dimensional root-finding problem. We take one element of the set of unknowns, for example, a_i^{peak1}, and set up a position-error function depending on this element:

$$ {}^{PosTriNegTri}p_i^{err1}\left(a_i^{peak1}\right) : \mathbb{R} \longrightarrow \mathbb{R} \,. \tag{4.33} $$

The derivation of ${}^{PosTriNegTri}p_i^{err1}\left(a_i^{peak1}\right)$ is done manually on the base of eqns. (4.21) – (4.32); since the description of the function is rather long, it can be found in App. B.1 (p. 189). Here, we assume ${}^{PosTriNegTri}p_i^{err1}\left(a_i^{peak1}\right)$ to be a transcendental function that can be set up on the base of the system of equations for $\Psi_i^{Step1} = PosTriNegTri$.

This results in a very simple and standard problem of curve sketching, which can be solved numerically. As we will see in the following, we are able to calculate an interval $\left[{}^{min}a_i^{peak1}, {}^{max}a_i^{peak1}\right]$ from which we know, that our desired value a_i^{peak1} lies within it. But as we will see, it may happen that there are up to two[4] valid values for a_i^{peak1} within this interval, and we are only interested in the minimum one, because it delivers *the* time-optimal solution with the minimum value for t_i^{min} (cf. Fig. 4.5). As a result, we also need the derivative of ${}^{PosTriNegTri}p_i^{err1}\left(a_i^{peak1}\right)$

$$ {}^{PosTriNegTri}p_i^{err1\prime}\left(a_i^{peak1}\right) = \frac{\mathrm{d}}{\mathrm{d}\,a_i^{peak1}}\ {}^{PosTriNegTri}p_i^{err1}\left(a_i^{peak1}\right) \tag{4.34} $$

[4] This number will be derived in detail in Chap. 5.1.1 (p. 73).

in order to solve an extremum problem for ${}^{PosTriNegTri}p_i^{err1}\left(a_i^{peak1}\right)$ and to determine a smaller interval $\left[{}^{min}a_i^{peak1}, {}^{max}a_i^{peak1}\right]$ that contains one, and only one, root. The full-length form of ${}^{PosTriNegTri}p_i^{err1\prime}\left(a_i^{peak1}\right)$ can be found in App. B.2 (p. 189).

Algorithm 4.1 shows the calculation steps to attain ${}^{min}a_i^{peak1}$ and ${}^{max}a_i^{peak1}$. Due to the previously executed decision tree for determining Ψ_i^{Step1}, we know that $A_i^{max} \geq A_i \geq 0$ holds (cf. *PosTriNegTri* acceleration profile in Fig. 4.4). The lower bound ${}^{min}a_i^{peak1}$ is calculated in lines $1-8$. Line 1 simply calculates the velocity $v^{(A_i \searrow 0)}$ we would reach if A_i was decreased to zero by applying J_i^{max}. If $v^{(A_i \searrow 0)}$ is greater than V_i^{trgt} (line 2), we can already set ${}^{min}a_i^{peak1}$ to A_i (line 3). Otherwise, we have to calculate an acceleration peak value with which V_i^{trgt} is exactly reached. Here, we only apply a positive triangle-shaped profile, that is, in line 7 we compute a simple acceleration profile consisting of two segments: Applying $+J_i^{max}$ until ${}^{min}a_i^{peak1}$ and then $-J_i^{max}$ until V_i^{trgt} is reached. The resulting value for ${}^{min}a_i^{peak1}$ constitutes the lower bound.

The lines $9-25$ of Algorithm 4.1 calculate the upper bound ${}^{max}a_i^{peak1}$. We first calculate two possible values. The first one, ${}^{max}a_i^{peak1,a}$ (line 11), is determined in order not to exceed $\pm A_i^{max}$ and the second one, ${}^{max}a_i^{peak1,v}$ (line 17), in order not to exceed $+V_i^{max}$. For the calculation of ${}^{max}a_i^{peak1,a}$, we compute an acceleration profile consisting of four segments as commented in line 10. If the resulting value is greater than A_i^{max}, we set ${}^{max}a_i^{peak1,a}$ to A_i^{max} in line 19. ${}^{max}a_i^{peak1,v}$ is calculated analogously. At the end (lines $21-25$) we set the actual value of ${}^{max}a_i^{peak1}$ to the lower value of ${}^{max}a_i^{peak1,a}$ or ${}^{max}a_i^{peak1,v}$, respectively.

As a result, we know that the function ${}^{PosTriNegTri}p_i^{err1}\left(a_i^{peak1}\right)$ is continuous between ${}^{min}a_i^{peak1}$ and ${}^{max}a_i^{peak1}$. If this interval contains two roots, we have to check for a local minimum or maximum by finding the (only) root ${}^{m}a_i^{peak1}$ of ${}^{PosTriNegTri}p_i^{err1\prime}\left(a_i^{peak1}\right)$ in $\left[{}^{min}a_i^{peak1}, {}^{max}a_i^{peak1}\right]$. The mathematical literature provides different methods for this standard problem. An overview of such numerical methods can be found in [73]. In our case, only inclusion methods come into consideration, and the most efficient inclusion method is the *Anderson-Björck-King method* [12, 73, 131]. Since this method may have robustness and thus real-time capability problems, it was adapted for this application by combining it with the simple bisection method. The details on the modified algorithm are presented and discussed in App. A (p. 185).

After the local minima and maxima of ${}^{PosTriNegTri}p_i^{err1}\left(a_i^{peak1}\right)$ are calculated, we set

$$ {}^{max}a_i^{peak1} \;:=\; {}^{m}a_i^{peak1} \;, \tag{4.35}$$

Algorithm 4.1 Calculate the interval limits for the desired root of ${}^{PosTriNegTri}p_i^{err1}\left(a_i^{peak1}\right)$.

Require: V_i^{trgt}, V_i^{max}, V_i, A_i^{max}, A_i, J_i^{max}, with $V_i^{max} \geq V_i$, $A_i^{max} \geq A_i \geq 0$, $J_i^{max} \geq 0$

Ensure: ${}^{min}a_i^{peak1}$, ${}^{max}a_i^{peak1}$

$v^{(A_i \searrow 0)} := V_i + \frac{(A_i)^2}{2J_i^{max}}$

if $\left(v^{(A_i \searrow 0)} > V_i^{trgt}\right)$ **then**

  ${}^{min}a_i^{peak1} := A_i$

else

  ▷ CALCULATE ${}^{min}a_i^{peak1}$ FOR THE PROFILE ◁

  ▷ $\left\{A_i \nearrow {}^{min}a_i^{peak1} \searrow 0, \text{ SO THAT } v = V_i^{trgt}\right\}$ ◁

  ${}^{min}a_i^{peak1} := \sqrt{\frac{(A_i)^2 + 4J_i^{max}\left(V_i^{trgt} - V_i\right)}{2}}$

end if

▷ CALCULATE ${}^{max}a_i^{peak1,a}$ FOR THE PROFILE ◁

▷ $\left\{A_i \nearrow {}^{max}a_i^{peak1,a} \searrow 0 \searrow -A_i^{max} \nearrow 0, \text{ SO THAT } v = V_i^{trgt}\right\}$ ◁

${}^{max}a_i^{peak1,a} := \frac{\sqrt{\left(J_i^{max}\right)^2\left((A_i)^2 + 2\left(A_i^{max}\right)^2\right) + \left(J_i^{max}\right)^3\left(V_i^{trgt} - V_i^{trgt}\right)}}{\sqrt{2}J_i^{max}}$

if ${}^{max}a_i^{peak1,a} > A_i^{max}$ **then**

  ${}^{max}a_i^{peak1,a} := A_i^{max}$

end if

▷ CALCULATE ${}^{max}a_i^{peak1,v}$ FOR THE PROFILE ◁

▷ $\left\{A_i \nearrow {}^{max}a_i^{peak1,v} \searrow 0, \text{ SO THAT } v = V_i^{max}\right\}$ ◁

${}^{max}a_i^{peak1,v} := \sqrt{\frac{(A_i)^2 + 4J_i^{max}\left(V_i^{max} - V_i\right)}{2}}$

if ${}^{max}a_i^{peak1,v} > A_i^{max}$ **then**

  ${}^{max}a_i^{peak1,v} := A_i^{max}$

end if

if ${}^{max}a_i^{peak1,v} < {}^{max}a_i^{peak1,a}$ **then**

  ${}^{max}a_i^{peak1} := {}^{max}a_i^{peak1,v}$

else

  ${}^{max}a_i^{peak1} := {}^{max}a_i^{peak1,a}$

end if

such that we finally obtain the desired interval $\left[{}^{min}a_i^{peak1}, {}^{max}a_i^{peak1}\right]$, which contains only one root (the desired value of a_i^{peak1}). We can again use the modified method of Anderson-Björck-King here. Once we know a_i^{peak1}, we know the first unknown variable of eqns. (4.21) – (4.32). The calculation of the other eleven unknowns is trivial, as is the calculation of the parameters of ${}^l\vec{m}_i(t)$ with $l \in \{1, \ldots, 4\}$ and ${}^l\vartheta_i$ with $l \in \{1, \ldots, 4\}$. These two steps can be developed in a straightforward way and are shown in App. B.3 (p. 190). Finally, with

$$\forall\, l \in \{1, \ldots, 4\}:$$

$${}^l\vec{m}_i(t) = \left({}^l p_i(t), {}^l v_i(t), {}^l a_i(t), {}^l j_i(t)\right) \tag{4.36}$$

$${}^l\mathcal{V}_i = \left\{{}^l\vartheta_i\right\} \qquad \text{with } {}^l\vartheta_i = \left[{}^l t_i, {}^{l+1}t_i\right] \tag{4.37}$$

the one-dimensional Type IV, Variant A trajectory $\mathcal{M}_i$ is completely described for one DOF at time instant T_i:

$$\mathcal{M}_i(t) = \left\{ \left({}^1\vec{m}_i(t), {}^1\mathcal{V}_i\right), \left({}^2\vec{m}_i(t), {}^2\mathcal{V}_i\right), \left({}^3\vec{m}_i(t), {}^3\mathcal{V}_i\right), \left({}^4\vec{m}_i(t), {}^4\mathcal{V}_i\right) \right\} . \tag{4.38}$$

According to eqns. (4.14) and (4.15), $\vec{M}_{i+1}$ can be calculated by finding the respective time interval $\hat{l}$, such that

$${}^{\hat{l}}t_i \leq \left(T_i + T^{cycle}\right) \leq {}^{\hat{l}+1}t_i \tag{4.39}$$

holds, and then we simply calculate the output values for lower-level control by

$$\vec{M}_{i+1} = {}^{\hat{l}}\vec{m}_i\left(T_i + T^{cycle}\right) . \tag{4.40}$$

These values are then used as set-points for lower-level motion controllers.

This part of the book exemplarily presented how to set up the *PosTri-NegTri* acceleration profile. The complexity of the OTG algorithm differs strongly, depending on the profile and its number of equations and unknown variables. For clarification: This intricately seeming procedure is only required because there is no analytical solution for the systems of equations for the acceleration profiles that can be computed absolutely robustly. All the chosen mathematical tools are of a very basic nature, but in this concrete case, these methods lead to sufficient results, and they perform efficiently and very robustly. Another advantage of the proposed method compared to the analytical one is that we directly obtain the desired solution for *the* time-optimal trajectory, and difficult selections of correct solutions can be omitted.

For a better understanding of this part of the book, we will apply this procedure to concrete values of $\vec{W}_i$ in the next part.

Example of Parameterizing the *PosTriNegTri* Acceleration Profile

This part explains the previously proposed procedure step by step by means of a concrete example. Let us take an arbitrary set of input values $\vec{W}_0$ at instant $T_0 = 0\,ms$:

$$\begin{aligned} P_0 &= -499\,mm & P_0^{trgt} &= -90\,mm \\ V_0 &= -335\,mm/s & V_0^{trgt} &= -347\,mm/s \\ A_0 &= 152\,mm/s^2 & V_0^{max} &= 985\,mm/s \\ A_0^{max} &= 972\,mm/s^2 & J_0^{max} &= 324\,mm/s^3\ . \end{aligned} \tag{4.41}$$

As the first step in the first control cycle at T_0, we apply the decision tree of Fig. 4.4 in order to determine the correct acceleration profile ${}^{r}\Psi^{Step1}$, which leads to *the* time-optimal trajectory. As a result we obtain the profile $\Psi_0^{Step1} = PosTriNegTri$, such that we can set up the function ${}^{PosTriNegTri}p_0^{err1}\left(a_0^{peak1}\right)$ corresponding to the respective system of equations (eqns. (4.21) – (4.32)). In the next step, we determine the interval $\left[{}^{min}a_0^{peak1}, {}^{max}a_0^{peak1}\right]$ in which the position-error function ${}^{PosTriNegTri}p_0^{err1}\left(a_0^{peak1}\right)$ is continuous, and in which the desired value of a_0^{peak1} is located. Applying Algorithm 4.1 to the concrete input values of eqn. (4.41) leads to

$${}^{min}a_0^{peak1} = 152.000\,mm/s^2 \qquad \text{and} \qquad {}^{max}a_0^{peak1} = 662.746\,mm/s^2\ . \tag{4.42}$$

We know that for $a_0^{peak1} \in \left[152.000\,mm/s^2, 662.746\,mm/s^2\right]$ up to two roots can be present. To illustrate this, the top part of Fig. 4.6 shows the progression of the position-error function ${}^{PosTriNegTri}p_0^{err1}\left(a_0^{peak1}\right)$ in this interval. Since the function values at the bounds have different signs, we already know that there is only one root in this interval. For other input parameters, for example, $\tilde{P}_0^{trgt} = -1090\,mm$, the abscissa in the top part of Fig. 4.6 would be shifted down by 1000 units (dash-dotted line)[5]. Then, two valid solutions for a_0^{peak1} could be found, and only the lower one would be relevant. For such a case, the derivative of the position-error function, as shown in the bottom part of Fig. 4.6, would be required to calculate the extremum of ${}^{PosTriNegTri}p_0^{err1}\left(a_0^{peak1}\right)$:

$${}^{m}a_0^{peak1} = 285.255\,mm/s^2\ , \tag{4.43}$$

which would subsequently be used as the upper interval limit in order to calculate the correct value of a_0^{peak1} (dashed lines in Fig. 4.6).

[5] For these input values, the *PosTriNegTri* acceleration profile would not be *the* time-optimal profile anymore. The *NegTriPosTri* acceleration profile would lead to *the* time-optimal solution here, but the *PosTriNegTri* profile also delivers two valid non-time-optimal solutions.

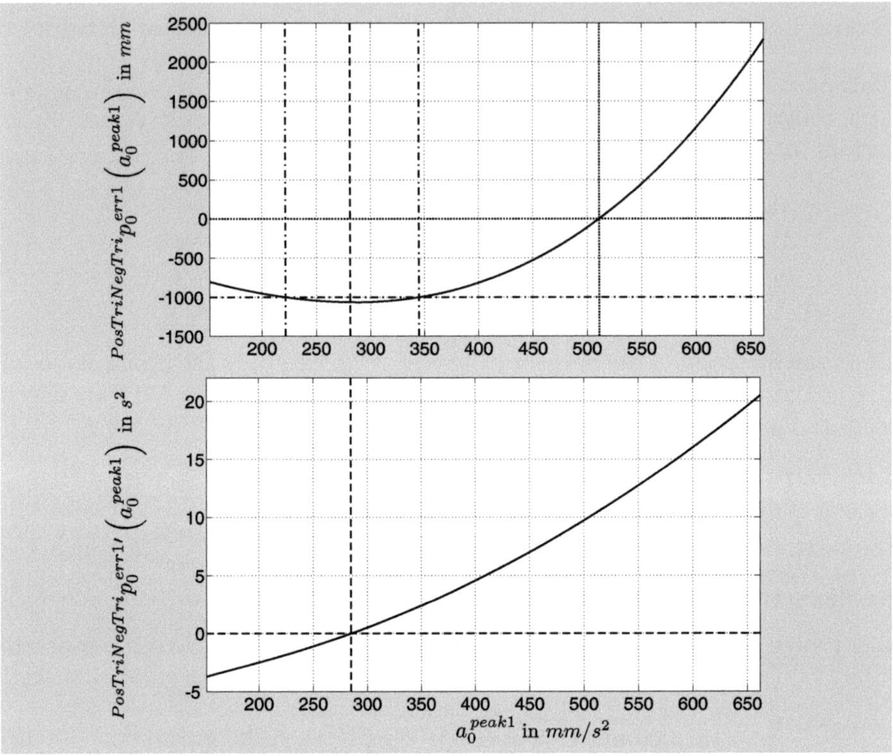

Fig. 4.6 Progressions of the position-error function and its derivative within the interval $[152.000mm/s^2, 662.746mm/s^2]$.

Regarding the initial input values of eqn. (4.41), we apply the modified Anderson-Björck-King method, and we obtain the desired value

$$a_0^{peak1} = 511.155\,mm/s^2 \tag{4.44}$$

as indicated by the dotted line in the top part of Fig. 4.6. By applying the assignments for the other eleven unknowns (App. B.3, p. 191, eqns. (B.13) – (B.23)), we obtain the following solution:

$$\begin{aligned}
t_i^{min} &= 5.795\,s & {}^2t_0 &= 1.109\,s \\
{}^3t_0 &= 2.686\,s & {}^4t_0 &= 4.240\,s \\
{}^2v_0 &= 32.555\,mm/s & {}^3v_0 &= 435.764\,mm/s \\
{}^4v_0 &= 44.382\,mm/s & {}^2p_0 &= -703.408\,mm \\
{}^3p_0 &= -227.969\,mm & {}^4p_0 &= 246.573\,mm \\
a_0^{peak1} &= 511.155\,mm/s^2 & a_0^{peak2} &= -503.603\,mm/s^2\ .
\end{aligned} \tag{4.45}$$

Afterwards, the parameters of $\mathcal{M}_0$ are computed by using eqns. (B.24) – (B.47). If we assume a cycle time T^{cycle} of one millisecond and refer to

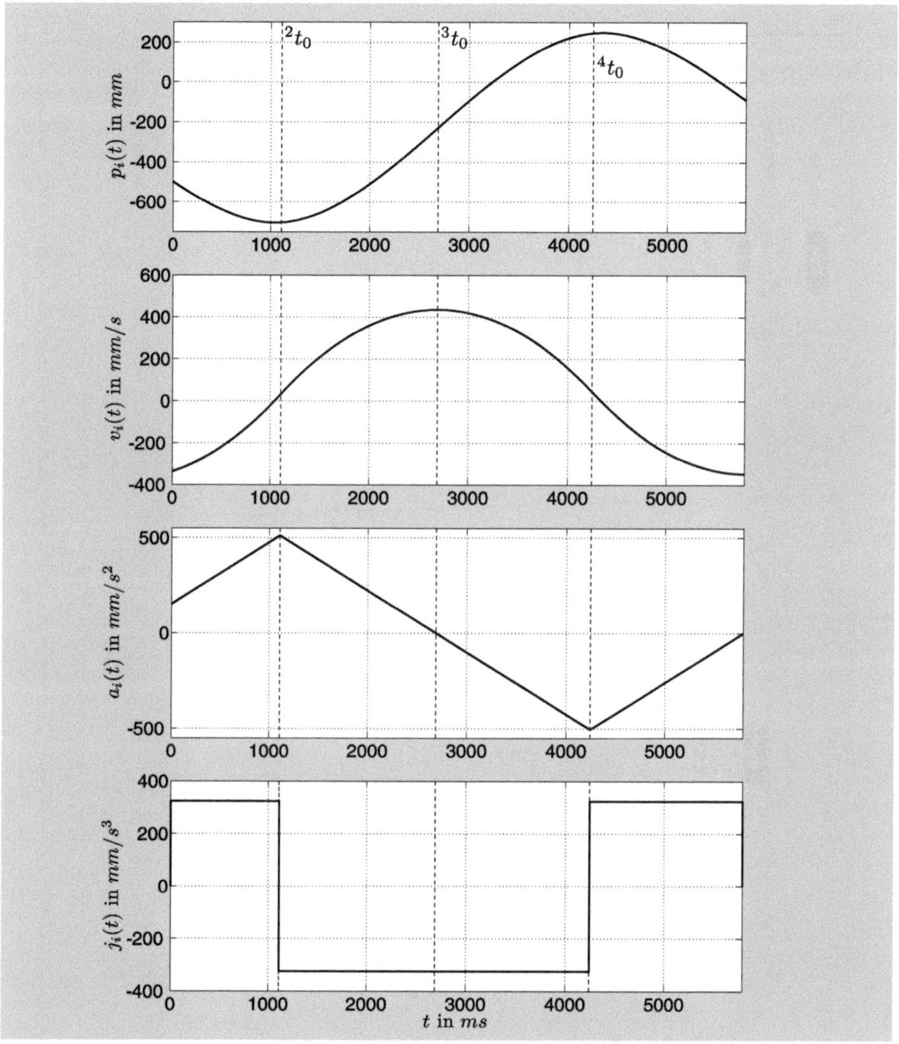

Fig. 4.7 Resulting trajectory $\mathcal{M}_i$ with $i \in \{0, \ldots, 5795\}$ for the given input values of eqn. (4.41). The dashed lines indicate the times $^2t_0$, $^3t_0$, and $^4t_0$ calculated in eqn. (4.45).

eqn. (4.39), we can find $\hat{l} = 1$. The output values $\vec{M}_1$ are subsequently calculated by eqn. (4.40):

$$\begin{aligned} P_1 &= -499.334\,mm & V_1 &= -334.847\,mm/s \\ A_1 &= 152.324\,mm/s^2 & J_1 &= 324\,mm/s^3 \ . \end{aligned} \qquad (4.46)$$

These values (cf. eqn. (4.41)) are used as set-points for lower-level control during the current control cycle. Hence, it would take us 5795 cycles until

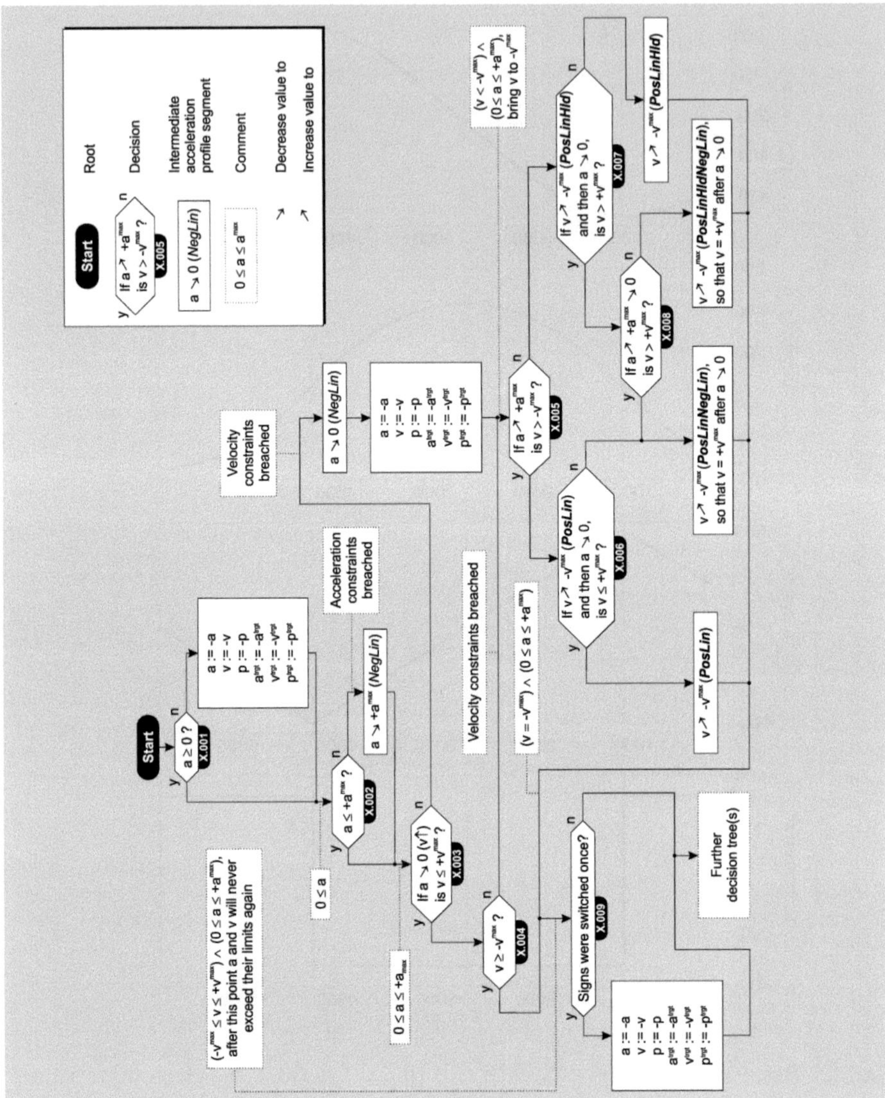

Fig. 4.8 Variant *B* decision tree, which is executed upstream to the one of Variant *A*. It is the task of this decision tree to bring all motion state values $\vec{M}_i$ into their limits $\vec{B}_i$ and to guarantee, that they remain within these bounds.

the desired target state of motion $\vec{M}_i^{trgt}$ is finally reached. If no sensor event occurs, the *consistency criterion* (Chap. 3.3, p. 42) must be fulfilled. Hence, if we execute the OTG algorithm again at $T_1 = 1\,ms$, we use the output values $\vec{M}_1$ from eqn. (4.46) as input values for the control cycle at $T_1 = 1\,ms$, and exactly the same trajectory must be calculated, that is, $\mathcal{M}_1$ must exactly fit

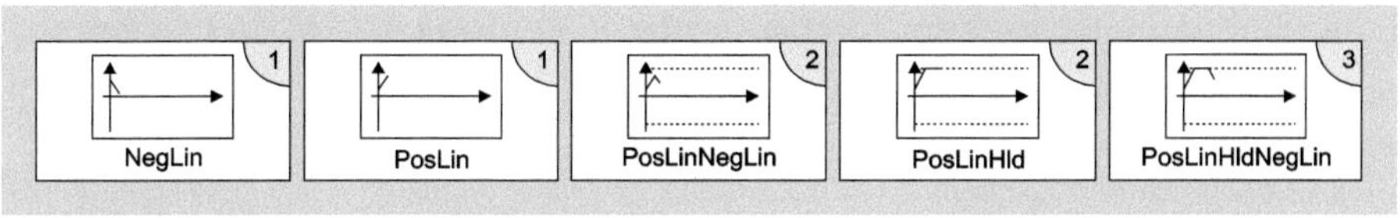

Fig. 4.9 Intermediate acceleration profiles for the Λ intermediate trajectory segments determined by the decision tree of Fig. 4.8.

into $\mathcal{M}_0$. The same holds, of course, for all following control cycles at instants T_i with $i \in \{2, \ldots, 5795\}$, as can be seen in Fig. 4.7, which depicts the resulting trajectory for the input values of eqn. (4.41). This is possible because we always compute *the* time-optimal trajectory, which transfers the system from $\vec{M}_0$ to $\vec{M}_0^{trgt}$ in *the* shortest possible time. That means for every motion state between $\vec{M}_0$ and $\vec{M}_0^{trgt}$ *the* time-optimal trajectory to reach $\vec{M}_0^{trgt}$ automatically fits into $\mathcal{M}_0$. The most important part to guarantee this essential property is the decision tree of Fig. 4.4. This tree as well as the acceleration profile set $\mathcal{P}_{Step1}$ must be complete and error-free, such that eqn. (4.13) holds (cf. p. 48).

Let us finally summarize this subsection briefly: The complete Type IV, Variant *A* OTG algorithm for one DOF was introduced and discussed. The acceleration profile *PosTriNegTri* belongs to the simplest ones and was chosen for a detailed demonstration. Finally, a very concrete example trajectory was calculated in order to improve the comprehension of the applied techniques.

4.2.2 Type IV, Variant B

Of much higher relevance than Variant *A* is Variant *B*, because it can cope with varying constraint values $\vec{B}_i$. Furthermore, Variant *B* fulfills the *criterion for motion constraints* (cf. Chap. 3.3, p. 41).

As generally described in Sec. 4.1.2 (p. 47), Λ additional trajectory segments ${}^{l}\vec{m}_i$ with $l \in \{1, \ldots, \Lambda\}$ with corresponding time intervals ${}^{l}\vartheta_i$ with $l \in \{1, \ldots, \Lambda\}$ (cf. eqns. (4.5) and (4.6)) have to be selected and parameterized. These intermediate trajectory segments are executed prior to the other $(L - \Lambda)$ segments of the Variant *A*.

Fig. 4.8 shows the respective decision tree in a compressed version. The execution of the intermediate trajectory segments shown in Fig. 4.9 is finished at ${}^{\Lambda+1}t_i$, and we have to assure, that

$$\begin{aligned} -V_i^{\,max} \leq\ & v_i\left({}^{\Lambda+1}t_i\right) \ \leq +V_i^{\,max} \quad \wedge \\ -A_i^{\,max} \leq\ & a_i\left({}^{(\Lambda+1)}t_i\right) \leq +A_i^{\,max} \end{aligned} \tag{4.47}$$

hold. Furthermore, we have to ensure that if we bring $a_i\left({}^{(\Lambda+1)}t_i\right)$ to zero by applying the maximum possible jerk J_i^{max}, the maximum velocity value of V_i^{max} is not exceeded again (neither positively nor negatively). Therefore, the plain condition

$$\left| v_i\left({}^{(\Lambda+1)}t_i\right) \pm \frac{\left(a_i\left({}^{(\Lambda+1)}t_i\right)\right)^2}{J_i^{max}} \right| \leq V_i^{max} \tag{4.48}$$

has to be fulfilled. The compressed version in Fig. 4.8 takes advantage of sign switchings. Since this tree is executed prior to all decision trees of OTG Types III-V and also prior to the further decision trees of the multi-dimensional case, the letter X has been chosen to replace the actual tree identifier. Decision *X.001* leads to a switching of signs for the initial and for the target state of motion if the current acceleration value A_i is negative. Hence, A_i is positive for decision *X.002*. This decision checks whether A_i^{max} is currently exceeded. If it is exceeded, we set up a first intermediate acceleration profile segment (*NegLin*), which brings A_i down to A_i^{max} by applying $-J_i^{max}$. The decisions *X.003* and *X.004* check whether V_i^{max} is positively or negatively exceeded. Since our current acceleration value is positive, decision *X.003* calculates the velocity value that we would obtain if we were to bring the acceleration value to zero (which increases the velocity value). If the resulting velocity is then greater than $+V_i^{max}$, we decrease the acceleration to zero by applying $-J_i^{max}$ again (*NegLin*), perform a switching of signs, and let the decisions *X.005* to *X.008* bring the velocity value into its bounds. Decision *X.004* only checks whether $-V_i^{max}$ is exceeded. If this is the case, we continue at decision *X.005*. For this decision, we know that the velocity is less than $-V_i^{max}$, and the acceleration is positive (no matter if the branch of decision *X.003* or *X.004* has been taken). If we would now increase the acceleration to $+A_i^{max}$, decision *X.005* checks whether the resulting velocity value is greater or less than $-V_i^{max}$. If it is less, we know that a simple acceleration increase brings the velocity value back into its limits, but we have to make sure, that it can remain within these. For this purpose, decision *X.006* checks whether $+V_i^{max}$ would be exceeded if we subsequently decreased the acceleration value to zero. If this is not the case (left branch), a simple *PosLin* profile segment, which applies $+J_i^{max}$, complies with the requirements of eqns. (4.47) and (4.48). Otherwise (right branch), we would increase the acceleration value to a certain peak value, and subsequently decrease it again, such that we would reach $+V_i^{max}$ exactly after the full decrease to zero (profile segment *PosLinNegLin*). The decisions *X.007* and *X.008* work analogously. In the last step, we have to re-switch the signs again if they have been switched before (decision *X.009*). Finally, we can assure that the conditions of eqns. (4.47) and (4.48) are fulfilled and will not be breached again, and we can continue with the decision trees of Variant A as presented in Sec. 4.2.1.

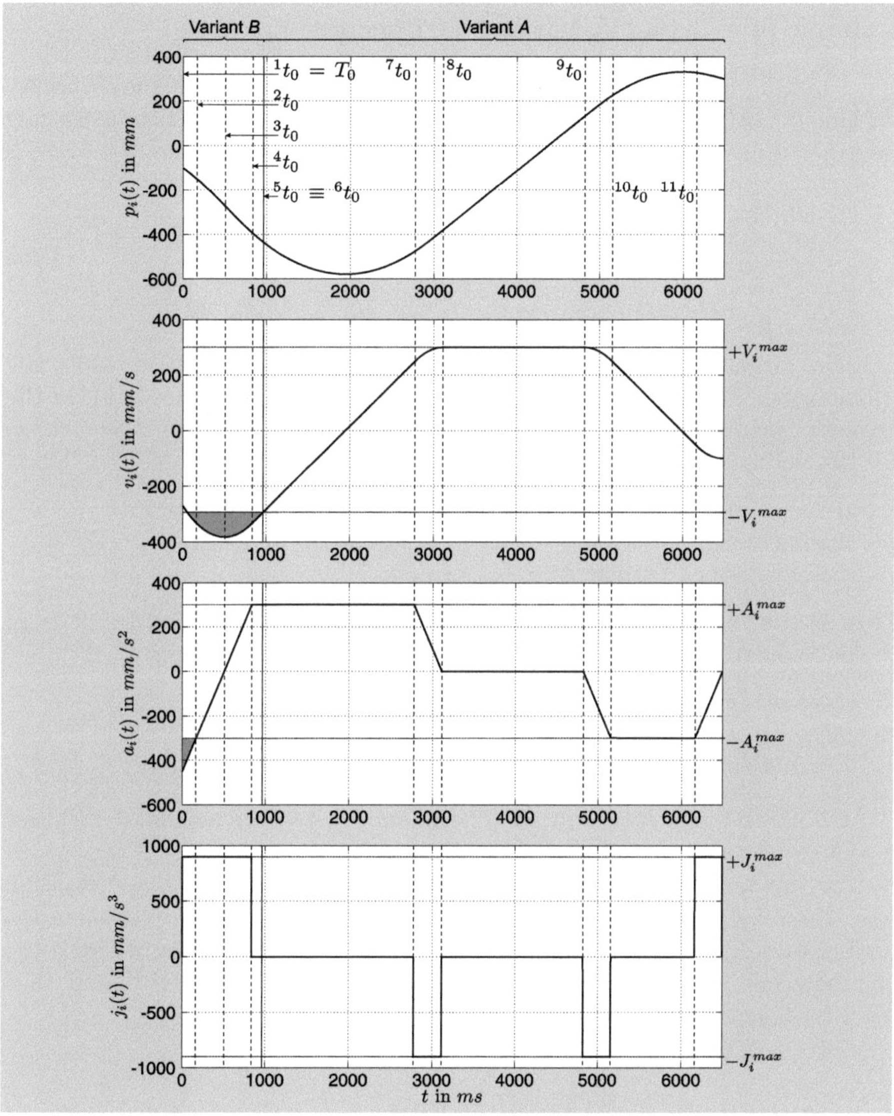

Fig. 4.10 Resulting Type IV, Variant B trajectory $\mathcal{M}_i$ with $i \in \{0, \ldots, N\}$ for the given input values of eqn. (4.49). The dashed lines indicate the bounds of the single trajectory segments, and the dotted lines indicate the motion constraints $\vec{B}_i$.

Once the Λ intermediate trajectory segments are determined, they have to be parameterized. This is done in the same way as for the acceleration profiles of Variant A (eqns. (4.21) – (4.32)), but the resulting systems of equations are of trivial nature and can be solved in a straightforward way without any numerical problems.

Example of a Type IV, Variant B Trajectory

For a better comprehension, we illustrate the functionality of the Type IV, Variant B OTG algorithm by means of a concrete example. Let us assume some given input values $\vec{W}_0$ at instant $T_0 = 0\,ms$:

$$\begin{aligned} P_0 &= -100\,mm & P_0^{trgt} &= 300\,mm \\ V_0 &= -270\,mm/s & V_0^{trgt} &= -100\,mm/s \\ A_0 &= -450\,mm/s^2 & V_0^{max} &= 300\,mm/s \\ A_0^{max} &= 300\,mm/s^2 & J_0^{max} &= 900\,mm/s^3\ . \end{aligned} \tag{4.49}$$

If the target motion state and the boundary values remain constant, the trajectory of Fig. 4.10 results from the input values $\vec{W}_0$ of eqn. (4.49). In the first step, we select intermediate trajectory segments by applying the decision tree from Fig. 4.8. Here, we would take the following path:

$X.001 \rightarrow$ Change of signs $\rightarrow X.002 \rightarrow NegLin \rightarrow X.003 \rightarrow NegLin$
$\rightarrow$ Change of signs $\rightarrow X.005 \rightarrow X.007 \rightarrow PosLinHld \rightarrow X.009$
$\rightarrow$ Decision tree of Variant A .

This results in $\Lambda = 4$ intermediate trajectory segments, which result from the decision tree of Variant B:

$NegLin \Longrightarrow PosLin$	One segment	$\left({}^{1}\vec{m}_0(t),\, {}^{1}\mathcal{V}_0\right)$
$NegLin \Longrightarrow PosLin$	One segment	$\left({}^{2}\vec{m}_0(t),\, {}^{2}\mathcal{V}_0\right)$
$PosLinHld$	Two segments	$\left({}^{3}\vec{m}_0(t),\, {}^{3}\mathcal{V}_0\right),\ \left({}^{4}\vec{m}_0(t),\, {}^{4}\mathcal{V}_0\right)$.

At T_0, both conditions, eqn. (4.47) and eqn. (4.48), are not fulfilled. These $\Lambda = 4$ trajectory segments lead to a new state of motion ${}^{\Lambda}\vec{m}_0({}^{(\Lambda+1)}t_0)$, which satisfies eqns. (4.47) and (4.48), and we can execute the Variant A decision tree. The result of this tree is the $\Psi_0^{Step1} = PosTrapZeroNegTrap$ acceleration profile. Finally, we obtain $L = 4 + 7 = 11$ trajectory segments, whereas the fifth segment is actually not existent, because ${}^{5}a_0\left({}^{5}t_0\right) = A_i^{max}$ and thus ${}^{5}t_0 \equiv {}^{6}t_0$ holds (cf. Fig. 4.10).

4.3 Summary and Applications

This chapter introduced the general OTG algorithm in Variant A and also in Variant B for one single DOF. Both variants fulfill the *time-optimality* and the *consistency criteria*, but only Variant B fulfills the *criterion for motion constraints* (cf. Chap. 3.3, p. 41). The presented approaches were exemplarily detailed by means of the Type IV on-line trajectory generation algorithm.

The three major scientific contributions of this chapter are the following:

Finite set of motion profiles
The basic idea of this work is, that a finite set $\mathcal{P}_{Step1}$ with R motion profiles ${}^{r}\Psi^{Step1}$ exist, from which one profile leads to *the* time-optimal trajectory.

Equation (4.13)

$$\bigcup_{r=1}^{R} {}^{r}\mathcal{D}_{Step1} \equiv \mathbb{R}^{\alpha}$$

Each motion profile ${}^{r}\Psi^{Step1}$ leads to a nonlinear system of equations, and each system of equations possesses a concrete input domain ${}^{r}\mathcal{D}_{Step1}$. The union of all domains is equivalent to the complete input space $\mathbb{R}^{\alpha}$ of the OTG algorithm (p. 48).

Decision trees

How can we describe and represent borders in the $\mathbb{R}^{\alpha}$ space? This is done by decision trees. The α input values of the OTG algorithm at instant T_i, that is, the elements of $\vec{W}_i$, are used to find a path through the tree in order to select the only correct motion profile $\Psi_i^{Step1} \in \mathcal{P}_{Step1}$, which leads to *the* time-optimal trajectory.

Before we extend the concept to multi-DOF systems, such that also the realization of the *criterion for time-synchronization* can be demonstrated, we should emphasize that we already developed an important and practically highly relevant byproduct for manufacturers of frequency inverters and servo drive systems.

All available products of the world's leading companies for electrical drive technologies (e.g.,[6] Bernecker + Rainer Industrie Elektronik GmbH [22], Lenze AG [159], Rockwell Automation [221], SEW-EURODRIVE [233], Siemens AG [242], YASKAWA Electric Corporation [281]) do not offer a general option for specifying and executing jerk-limited trajectories. Commonly only Type II trajectories with unlimited jerks are generated. All these manufacturers are aware of the problem, and all of them know that jerk limitation is indispensable for a wide field of applications. Nevertheless, usually only a pseudo jerk limitation is offered, which does not work in general but only in special modes.

Let us exemplarily take the manual of a *Lenze 9400 StateLine*™ frequency inverter [158], which is an up-to-date, high-quality, and high-performance device, to represent the state of the art in technology in this field. The first possibility to enable a kind of jerk limitation is using a current set-point filter in the velocity control loop at a certain frequency [Hz], with a certain width [Hz], and with a certain depth [dB]. Of course, this simple idea limits the jerk of an executed trajectory, but the jerk limitation depends on the currently executed motion and cannot be deterministically planned. Furthermore—depending on the application—the filtering effect might be undesired, for example, during a high-performance motion that requires low tracking errors.

A second opportunity for jerk limitation is the option of specifying a so-called "S-ramp time." This time is comparable to a real jerk limitation and is directly related to J_i^{max}, but this works only if the drive is controlled manually,

[6] In alphabetical order.

and it is required that the initial speed and the initial acceleration are zero. Hence, this is also not practicable for common operation.

Even a simple stop, that is, a controlled decrease of the current velocity to zero, at an unforeseen time instant, leads to an infinite jerk. To solve this problem, the Type III OTG algorithm would already satisfy the demands. This chapter already presented the more advanced Type IV OTG algorithm, which would not only enable the performing of a jerk-limited stop motion but also a jerk-limited motion to a certain velocity unequal to zero.

Apart from the benefit of exerting such "smooth" motions, it would additionally become possible to consider the rotor response time by specifying a maximum jerk for a concrete motor. In turn, this leads to a better control behavior of the whole servo system.

The Type IV, Variant *B* OTG algorithm would overcome many of the problems the aforementioned companies deal with. The algorithm works robustly, is real-time capable, and its interface (Fig. 4.2, p. 50) is very simple, such that an integration into existing systems could be realized without very much effort. Additionally, the usage as submodule in a hybrid switched-system controller (cf. Fig. 2.3, p. 24) is, of course, possible.

Chapter 5
Solution in Multi-dimensional Space

The previous chapter introduced the concept of OTG for one single DOF. This chapter extends the proposed class of algorithms to the multi-dimensional case, and we will see how to fulfill the *time-synchronization criterion.* The OTG algorithm in multi-dimensional space consists of three basic steps; the first step is based on the methodologies introduced in Chap. 4. Finally, this concept will be suitable as a submodule in a hybrid switched-system controller for a robotic system with multiple DOFs. As in the previous chapter, we first introduce the general OTG algorithm, and subsequently, the Type IV OTG algorithm is exemplarily outlined.

5.1 General Variant A Algorithm for On-Line Trajectory Generation

This section describes the Variant A OTG algorithm for multiple DOFs generally, such that it is applicable to all types and variants of OTG. In contrast to the last chapter, which considered only one DOF, we consider K DOFs here. The input values of the algorithm are represented by the matrix $\mathbf{W}_i$ and the output values by the matrix $\mathbf{M}_{i+1}$ (cf. Fig. 3.3, p. 37).

We assume that the algorithm is executed at the time instant T_i; all types of OTG require the same three algorithmic steps:

Step 1
Calculation of the minimum possible synchronization time t_i^{sync}.
Step 2
Synchronization of all selected DOFs to t_i^{sync} and calculation of all trajectory parameters $\mathcal{M}_i$.
Step 3
Calculation of all output values $\mathbf{M}_{i+1}$ based on $\mathcal{M}_i$.

These three steps are depicted in Fig. 5.1, which shows the overall structogram of the OTG algorithm. The following three subsections describe each step in detail.

T. Kröger: On-Line Trajectory Generation in Robotic Systems, STAR 58, pp. 69–97.
springerlink.com

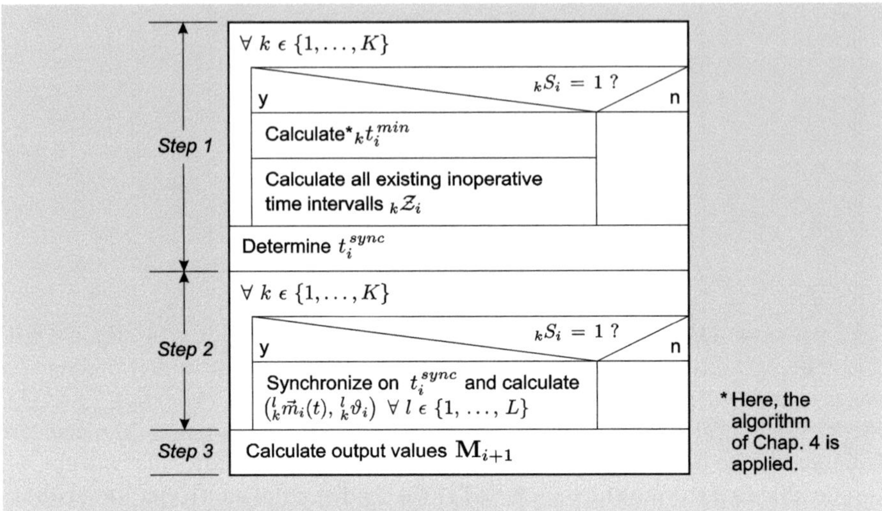

Fig. 5.1 Nassi-Shneiderman structogram of the OTG algorithm.

5.1.1 *Step 1: Calculating the Synchronization Time t_i^{sync}*

This step is the most complex one, although it only computes the synchronization time t_i^{sync}, that is, one scalar value. It is a function

$$f:\ \mathbb{R}^{\alpha K} \times \mathbb{B}^K \longrightarrow \mathbb{R}\,. \tag{5.1}$$

As can be seen in Fig. 5.1, *Step 1* can be subdivided into three parts: the individual calculation of the minimum execution time ${}_kt_i^{min}$ for each selected DOF k; the calculation of the set of possibly existing inoperative time intervals ${}_k\mathcal{Z}_i$, in which a selected DOF k cannot be synchronized; and finally the determination of t_i^{sync}.

Minimum Execution Times

At this particular point, we benefit from the last chapter, in which we calculated complete trajectories $\mathcal{M}_i$ for single DOFs. Here we are interested in one particular parameter of $\mathcal{M}_i$: We need the minimum execution time ${}_kt_i^{min}$ for each selected DOF k, which is required to transfer the DOF k from its initial state of motion ${}_k\vec{M}_i$ to its target state of motion ${}_k\vec{M}_i^{trgt}$. Hence, we execute the algorithm for the one-dimensional case of Chap. 4 only until we have calculated ${}_kt_i^{min}$. This is done for every single selected DOF, such that we attain up to K times ${}_kt_i^{min}$ with $k \in \{1,\ldots,K\}$.

Inoperative Time Intervals

After the minimum execution time ${}_kt_i^{min}$ for a DOF k has been calculated, we have to check, whether it is possible to execute the trajectory for this DOF within *any* time $t > {}_kt_i^{min}$. If this is the case, no inoperative time intervals ${}_k\mathcal{Z}_i = \{\}$ exist; depending on the type of OTG, ${}_k\mathcal{Z}_i$ may contain up to $Z = 3$ time intervals, in which a selected DOF k cannot be synchronized. Referring to Table 3.1 (p. 39) this property can be expressed by

$$Z = \alpha - 2\beta - 1 \ , \tag{5.2}$$

such that the following values of Z appear:

$$\begin{aligned} \textit{Types I, III, VI}: Z &= 0 \\ \textit{Types II, IV, VII}: Z &= 1 \\ \textit{Types V, VIII}: Z &= 2 \\ \textit{Type IX}: Z &= 3 \ . \end{aligned}$$

The reason for this characteristic is simple: For each of the target motion state parameters ${}_kV_i^{trgt}$, ${}_kA_i^{trgt}$, and ${}_kJ_i^{trgt}$, there may be one interval in which one of these values cannot be reached. Hence, eqn. (5.2) is exactly in accordance with Table 3.1.

A single element of a set ${}_k\mathcal{Z}_i$ is denoted by

$${}_k^z\zeta_i = \left[{}_k^zt_i^{begin}, {}_k^zt_i^{end}\right], \text{ with } z \in \{1, \ldots, Z\} \ , \tag{5.3}$$

such that we obtain

$${}_k\mathcal{Z}_i = \left\{{}_k^1\zeta_i, \ldots, {}_k^Z\zeta_i\right\} \tag{5.4}$$

if $Z > 0$. It is clear, that

$${}_kt_i^{min} \leq {}_k^zt_i^{begin} \leq {}_k^zt_i^{end} \quad \forall \ (z, k) \in \{1, \ldots, Z\} \times \{1, \ldots, K\} \tag{5.5}$$

naturally holds.

To explain the origin of these inoperative time intervals, Fig. 5.2 illustrates an example of a simple Type II trajectory for one single DOF k with simple bang-bang characteristics and one inoperative time interval ${}_k^1\zeta_i$. At $T_0 = 0\,ms$, let us assume the following input values for the Type II trajectory of Fig. 5.2:

$$\left.\begin{array}{rcl} \left.\begin{array}{rcl} {}_kP_0 &=& 50\,mm \\ {}_kV_0 &=& 80\,mm/s \end{array}\right\} {}_k\vec{M}_0 \\ \left.\begin{array}{rcl} {}_kP_0^{trgt} &=& 300\,mm \\ {}_kV_0^{trgt} &=& 70\,mm/s \end{array}\right\} {}_k\vec{M}_0^{trgt} \\ \left.\begin{array}{rcl} {}_kV_0^{max} &=& 200\,mm/s \\ {}_kA_0^{max} &=& 20\,mm/s^2 \end{array}\right\} {}_k\vec{B}_0 \\ {}_kS_0 \ = \ 1 \qquad\qquad \end{array}\right\} {}_k\vec{W}_0 \ . \tag{5.6}$$

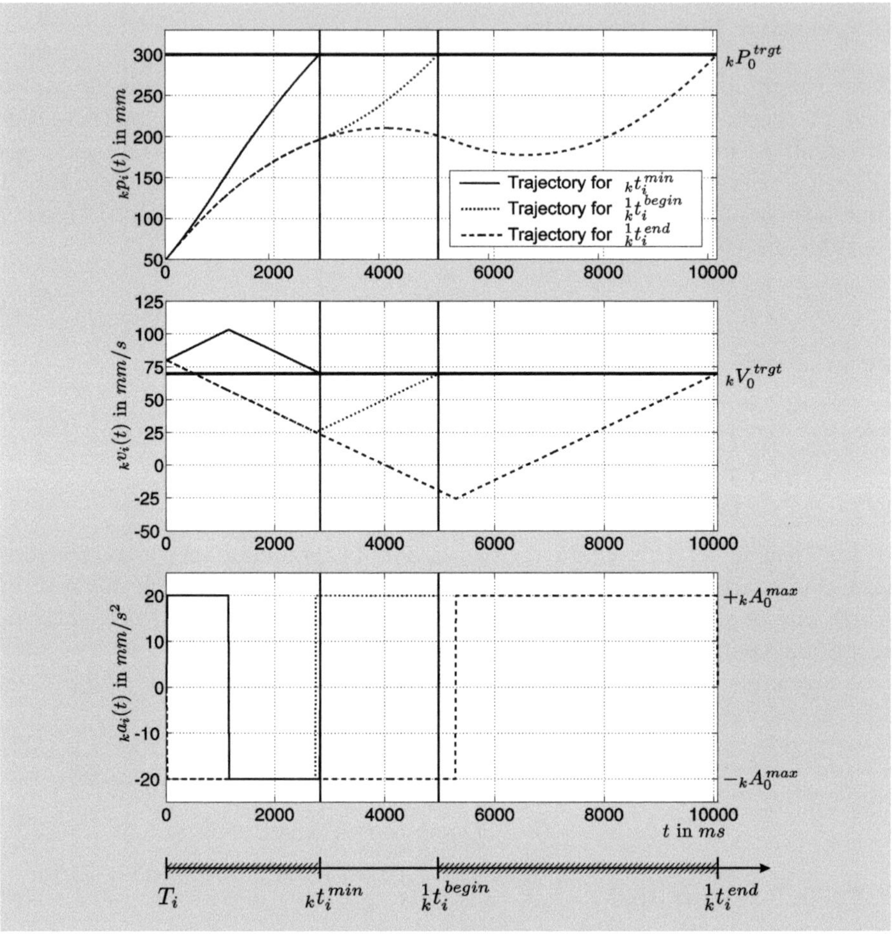

Fig. 5.2 Example of an inoperative time interval $^1_k\zeta_i$ for one translational DOF k calculated by a Type II on-line trajectory generation algorithm. The interval limits are $^1_kt_i^{begin} = 4950ms$ and $^1_kt_i^{end} = 10050ms$. During all three trajectories, either $+_kA_0^{max} = 20mm/s^2$ or $-_kA_0^{max} = -20mm/s^2$ is applied as value for $_ka_i(t)$ (cf. eqn. (5.6)). The desired target state of motion $_k\vec{M}_0^{trgt}$ can only be reached if $t_i^{sync} \in \left[{_kt_i^{min}}, {^1_kt_i^{begin}}\right]$ or $t_i^{sync} \geq {^1_kt_i^{end}}$.

The time-optimal case for transferring the DOF from $_k\vec{M}_0$ to $_k\vec{M}_0^{trgt}$ would take $_kt_i^{min} = 2820ms$ by applying a *PosTri* velocity profile (solid line). If other DOFs require more time, this DOF k can also reach $_k\vec{M}_i^{trgt}$ after $^1_kt_i^{begin} = 4950ms$ (or any time between $_kt_i^{min}$ and $^1_kt_i^{begin}$) by applying another velocity profile. But it is impossible to transfer DOF k to $_k\vec{M}_i^{trgt}$ in a time $^1_kt_i^{begin} < t < {^1_kt_i^{end}}$ with $^1_kt_i^{end} = 10050ms$. If we decreased the velocity to less than $25.5\,mm/s$ (dotted line in Fig. 5.2), there would not be enough space to accelerate again to ${}_kV_0^{trgt} = 70mm/s$. In such a case, the respective DOF has to decrease its velocity until the velocity is negative and the distance to ${}_kP_0^{trgt}$ equals the required acceleration distance (dashed line). The execution of this second *NegTri* velocity profile would be finished at ${}_k^1t_i^{end}$. For all times $t \geq {}_k^1t_i^{end}$ the synchronized motion of DOF k is possible again. As a result for the input values of eqn. (5.6), an inoperative time interval of

$$ {}_k^1\zeta_i = [4950ms, 10050ms] \tag{5.7}$$

exists.

It is *absolutely essential* that the OTG algorithm is able to provide a solution for *any* set of input values $\mathbf{W}_i \in \mathbb{R}^{\alpha K} \times \mathbb{B}^K$. If this cannot be guaranteed, the concept will be incomplete, unsafe, and thus practically irrelevant. We must be able to specify a set of times for each selected DOF k

$$ {}_k\mathcal{I}_i = \left\{ {}_kt_i^{min}, {}_k^1t_i^{begin}, {}_k^1t_i^{end}, \ldots, {}_k^Zt_i^{begin}, {}_k^Zt_i^{end} \right\} , \tag{5.8}$$

which contains ${}_kt_i^{min}$ and all interval limits of the set ${}_k\mathcal{Z}_i$. It belongs to the nature of motion profiles of $\mathcal{P}_{Step1}$ that one particular profile can deliver up to two valid solutions:

- If the system of equations of a motion profile ${}^r\Psi^{Step1}$ (e.g., eqns. (4.21) – (4.32), p. 53) delivers only one valid solution, this solution specifies *one element* of the set of times ${}_k\mathcal{I}_i$.
- If two valid solutions are available for a system of equations, *two neighbored elements* of the set ${}_k\mathcal{I}_i$ can be specified.

More than two solutions are not possible, because then at least one element of ${}_k\vec{M}_i^{trgt}$ cannot be reached. Let us explain this important property by means of two examples.

Example 5.1. *The system of equations for the PosTri velocity profile of Fig. 5.2 delivers one (and only one) valid solution, which yields the minimum execution time* ${}_kt_i^{min}$*. It is impossible, that this profile can deliver a second valid solution for the given input values* ${}_k\vec{W}_i$ *of eqn. (5.6), because all velocity peak values that are greater than the one from the first solution would lead to an irreversible overshooting of* ${}_kP_i^{trgt}$*. The system of equations for the NegTri velocity profile, however, delivers two valid solutions. The first one contains the left bound* ${}_k^1t_i^{begin}$ *of the inoperative time interval, and the second one contains the respective right bound* ${}_k^1t_i^{end}$*. Apart from these two solutions, no further valid solutions are possible, because* ${}_kP_i^{trgt}$ *could not be reached then.*

Example 5.2. *This example builds the bridge to Chap. 4.2.1 (p. 55), where we already claimed, that a particular motion profile can deliver up to two valid solutions to set up a trajectory. From the example of eqn. (4.41) (p. 59), we learned that the PosTriNegTri acceleration profile leads to the time-optimal* *motion for that DOF. The plot of the position-error function of Fig. 4.6 (p. 60) illustrates how the solution for the value of* a_0^{peak1} *is calculated. But the figure contains a second piece of information: If we subtract* $1000\,mm$ *from the original value of the desired target position* P_0^{trgt}*, two solutions can be found for the PosTriNegTri acceleration profile. If applied as DOF k of a multi-DOF system, these two solutions would lead to the bounds* ${}_k^1t_i^{begin}$ *and* ${}_k^1t_i^{end}$ *of the inoperative time interval* ${}_k^1\zeta_i$*. Further valid solutions are not possible, because* ${}_kP_i^{trgt}$ *would not be reached then.*

Whether a solution of a system of equations for a motion profile ${}^r\Psi^{Step1}$ contributes to the elements of the set ${}_k\mathcal{I}_i$, is specified by further decision trees. We need $2Z$ further decision trees — that is, two for each possible inoperative interval ${}_k^z\zeta_i$ — in order to obtain a complete set ${}_k\mathcal{I}_i$ for each selected DOF k. If one or more inoperative time intervals ${}_k^z\zeta_i$ exist for a selected DOF k, each decision tree determines a certain motion profile whose system of equations leads to a solution that contains a lower or an upper bound of an inoperative time interval ${}_k^z\zeta_i$.

The quintessence of this paragraph is that $(2Z+1)$ decision trees are required to calculate the minimum execution times ${}_kt_i^{min}$ and the set of all inoperative time intervals ${}_k\mathcal{Z}_i$ for all selected DOFs $k \in \{1, \ldots, K\}$. All these values are subsequently required for the determination of the minimum synchronization time t_i^{sync}.

Determining t_i^{sync}

After the minimum execution times and all existing inoperative time intervals have been calculated for all selected DOFs, t_i^{sync} can be determined easily. First, the greatest time of all minimum execution times is determined. t_i^{sync} cannot be less than this value. In the second step, we have to consider that t_i^{sync} must not be part of any inoperative time interval ${}_k^z\zeta_i$ with $(z,k) \in \{1, \ldots, Z\} \times \{1, \ldots, K\}$. Fig. 5.3 illustrates an example with four DOFs, where $t_i^{sync} = {}_2^1t_i^{end}$ is determined as the minimum possible synchronization time.

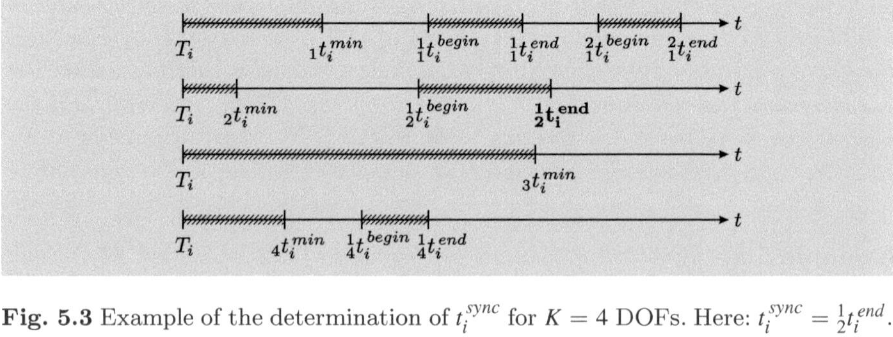

Fig. 5.3 Example of the determination of t_i^{sync} for $K = 4$ DOFs. Here: $t_i^{sync} = {}_2^1t_i^{end}$.

5.1.2 Step 2: Synchronization

Due to Step 1, we know that every selected DOF is able to reach its target state of motion ${}_k\vec{M}_i^{trgt}$ at t_i^{sync}. It is clear that for all DOFs k, whose minimum execution time ${}_kt_i^{min}$ is less than t_i^{sync}, in principal an infinite number of solutions can be found to parameterize a trajectory that transfers a respective DOF from ${}_k\vec{M}_i$ to ${}_k\vec{M}_i^{trgt}$ in t_i^{sync}. For the proposed OTG algorithm it is important to work deterministically, such that we are interested in one particular solution for every selected DOF. Thus, we have to find the same solution for the same input values, and furthermore, a continuous change of input values must result in a continuous and jump-free change of the trajectory parameters $\mathcal{M}_i$. To cope with these demands, we apply an optimization method to a certain trajectory property. Depending on the type of OTG, one of the following criteria has to be fulfilled for each selected DOF $k \in \{1,\ldots,K\}$:

$$\text{Types I--II}: \quad \int_{T_i}^{t_i^{sync}} \left|{}_ka_i(t)\right| dt \quad \rightarrow \min \tag{5.9}$$

$$\text{Types III--V}: \quad \int_{T_i}^{t_i^{sync}} \left|{}_kj_i(t)\right| dt \quad \rightarrow \min \tag{5.10}$$

$$\text{Types VI--IX}: \quad \int_{T_i}^{t_i^{sync}} \left|{}_kd_i(t)\right| dt \quad \rightarrow \min\ . \tag{5.11}$$

Equations (5.9) – (5.11) constitute the simplest case, which leads to simple rectangular profiles for the acceleration (eqn. (5.9)), the jerk (eqn. (5.10)), or the derivative of jerk (eqn. (5.11)). This is an important simplification of the problem, because this way we can apply the same kind of motion profiles in Step 2, which we already know from Step 1.

Remark 5.1. *In order to explain, why the minimization criteria of eqns. (5.9) – (5.11) leads to rectangular motion profiles more comprehensively, let us consider the following sample input values for Step 2 of the simple Type I OTG algorithm at instant* $T_0 = 0\,ms$*:*

$$\left.\begin{array}{rcl} \left.\begin{array}{rcl} {}_kP_0 &=& -200\,mm \\ {}_kV_0 &=& 0\,mm/s \end{array}\right\} {}_k\vec{M}_0 \\ \left.{}_kP_0^{trgt} = 300\,mm\right\} {}_k\vec{M}_0^{trgt} \\ \left.\begin{array}{rcl} {}_kV_0^{max} &=& 500\,mm/s \\ {}_kA_0^{max} &=& 400\,mm/s^2 \end{array}\right\} {}_k\vec{B}_0 \\ {}_kS_0 = 1 \end{array}\right\} {}_k\vec{W}_0$$

$$t_0^{sync} = 2830\,ms\ . \tag{5.12}$$

The minimum execution time for this set of input parameters ${}_k\vec{W}_0$ for DOF k is ${}_kt_0^{min} = 2.237\,ms$, but since the minimum execution of at least one other selected DOF is assumed to be greater, we assume a synchronization time of $t_0^{sync} = 2.830\,ms$ as the resulting value from Step 1 of the Type I OTG algorithm. As a result, there are infinite possibilities for Step 2 to set up a trajectory for DOF k; three of them are shown in Fig. 5.4. The trajectory represented by the solid line fulfills eqn. (5.9). This criterium can only be fulfilled if only $\pm_k A_0^{max}$ or zero acceleration are utilized. For the OTG Types III – V, only $\pm_k J_0^{max}$ or a jerk value of zero are allowed to be utilized for all $k \in \{1, \ldots, K\}$, and for the OTG Types VI – IX the same holds for the maximum derivative of jerk, ${}_kD_0^{max}$.

$\mathbf{W}_i$ and t_i^{sync} act as input parameters for this second step, in which a similar procedure as in the first part of Step 1 is applied. The key idea — that there is a *finite* set of motion profiles to set up the final motion trajectories — is employed again. The elements of this set, $\mathcal{P}_{Step2}$, differ from the ones of Step 1, because t_i^{sync} is considered as an additional input value. These motion profiles are denoted as

$$\mathcal{P}_{Step2} = \left\{ {}^1\Psi^{Step2}, \ldots, {}^s\Psi^{Step2}, \ldots, {}^S\Psi^{Step2} \right\}, \tag{5.13}$$

where S denotes the number of elements in $\mathcal{P}_{Step2}$ and depends on the type of OTG. According to Table 3.1 (p. 39), a decision tree that acts as function

$$f: \left(\left(\mathbb{R}^{\alpha+1} \right) \setminus \mathcal{H} \right) \longrightarrow \mathcal{P}_{Step2} \tag{5.14}$$

is required, which determines a motion profile ${}^s\Psi^{Step2}$ for each selected DOF k. The meaning of $\mathcal{H}$ will be described below. Compared to Step 1, Step 2 always requires only one decision tree, which is executed once for every selected DOF and control cycle. But compared to Step 1, there are two significant differences:

1. While in Step 1 more than one system of equations may achieve a valid solution (cf. Fig. 4.6, p. 60), in Step 2 — that is responsible for the actual calculation of the trajectory parameters of $\mathcal{M}_i$ — only one element of $\mathcal{P}_{Step2}$ leads to a solvable system of equations. This fact is rooted in the nature of the Step 2 motion profiles and leads to a deterministic behavior, that is, there can always be only one possible trajectory, which transfers a selected DOF k from ${}_k\vec{M}_i$ to ${}_k\vec{M}_i^{trgt}$ in t_i^{sync}.
2. If we denote the input domain of the system of equations that corresponds to the motion profile ${}^s\Psi^{Step2}$ as

$$ {}^s\mathcal{D}_{Step2} \subset \mathbb{R}^{\alpha+1}, \tag{5.15}$$

   we do *not* have such an equivalence as described by eqn. (4.13) for Step 1 (p. 48):

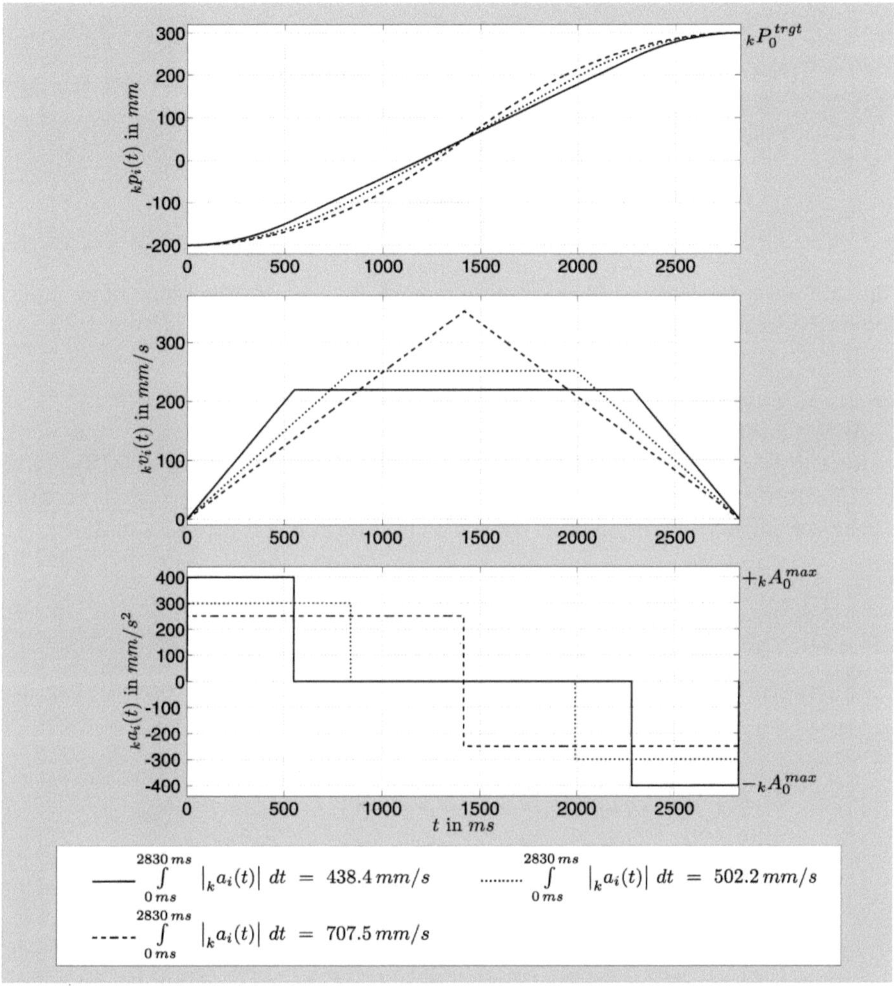

Fig. 5.4 Three different, valid Type I trajectories for the input values of eqn. (5.12). The solid line represents the trajectory that fulfills the minimization criterium of eqn. (5.9).

$$\bigcup_{s=1}^{S} {}^{s}\mathcal{D}_{Step2} \not\equiv \mathbb{R}^{\alpha+1} . \qquad (5.16)$$

Due to the inoperative time intervals ${}_k\mathcal{Z}_i$ for each selected DOF k, the $(\alpha+1)$-dimensional space contains *holes*, in which none of the systems of equations that correspond to the motion profiles of $\mathcal{P}_{Step2}$ is solvable, and in which the decision tree of eqn. (5.14) does not deliver a solution. These holes in the $(\alpha+1)$-dimensional space are represented by the set

$$\mathcal{H} \subset \mathbb{R}^{\alpha+1} . \qquad (5.17)$$

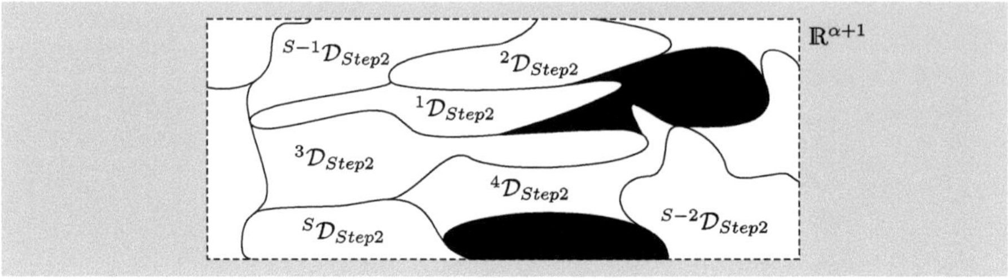

Fig. 5.5 Two-dimensional illustration of eqn. (5.18), that is, the union of all input domains $^{s}\mathcal{D}_{Step2}$ of the systems of equations for the motion profiles of Step 2, $\mathcal{P}_{Step2}$. The black areas represent $\mathcal{H}$.

As a result, the union of all input domains of the systems of equations for the Step 2 motion profiles $\mathcal{P}_{Step2}$ can be described in eqn. (5.18). This equation is illustrated by Fig. 5.5. We can also logically conclude here that the set of holes $\mathcal{H}$ is disjunctive with the union of all input domains:

$$\bigcup_{s=1}^{S} {^{s}\mathcal{D}_{Step2}} = \left(\mathbb{R}^{\alpha+1}\right) \setminus \mathcal{H} \quad \Longrightarrow \quad \bigcup_{s=1}^{S} \left({^{s}\mathcal{D}_{Step2}} \cap \mathcal{H}\right) = \{\ \}\,. \tag{5.18}$$

Furthermore, it is an important property that the union of pairwise intersections of all Step 2 input domains

$$\mathcal{S} = \bigcup_{s=1}^{S} \left\{ {^{s}\mathcal{D}_{Step2}} \cap {^{u}\mathcal{D}_{Step2}} \ \forall\ u \in \{1,\ldots,S\} \middle| u \neq s \right\} \tag{5.19}$$

constitutes α-dimensional hyperplanes in the $(\alpha+1)$-dimensional space. These hyperplanes are described in detail by the decision tree of eqn. (5.14). The fact that $\mathcal{S}$ represents only hyperplanes in the input space of Step 2 is essential for the deterministic behavior of the OTG algorithm, because this way there is only one possible motion profile for any set of input values. If the $(\alpha+1)$ input values for a selected DOF k are an element of $\mathcal{S}$, two or even more motion profiles will lead to a solvable system of equations, but the solution will be *exactly* the same for all profiles. This fact emphasizes the *overall consistency* of the proposed concept.

Let us now answer the question: How can we describe the set of holes $\mathcal{H}$? This is an important question, because we have to prevent form entering such a hole. Entering a hole would lead to an unsolvable problem, and no output values $\mathbf{M}_{i+1}$ could be calculated.

The $(\alpha+1)$-dimensional space for one DOF k is spanned by the first α elements of $_{k}\vec{W}_{i}$ and by the synchronization time t_i^{sync}. The values of $_{k}\vec{W}_{i}$ can be arbitrary, as described in the problem formulation in Chap. 3.3 (p. 40). Hence, the only parameter that can be responsible for $\mathcal{H}$, is t_i^{sync}, which was determined in Step 1. As described there, we need $2Z$ decision trees to determine the limits of all inoperative time intervals ${}^{z}_{k}\zeta_i$ with $(z,k) \in \{1,\ldots,Z\} \times \{1,\ldots,K\}$. These $2Z$ decision trees exactly describe the set of holes $\mathcal{H}$, because we mask the codomain of t_i^{sync} and thus the $(\alpha+1)$-dimensional input space.

An important insight at this point is that all $(2Z+1)$ decision trees of Step 1 and the decision tree of Step 2 must *exactly* fit each other. It is comparable to a multi-dimensional puzzle game whose parts have to fit each other absolutely exactly, and if this were not the case, the game would not come to an end. The same holds for the OTG algorithm: If the $(2Z+1)$ Step 1 decision trees do not fit each other, the algorithm will be erroneous, and there will exist input values $\mathbf{W}_i$, to which no output values can be calculated.

Once the correct motion profile ${}_{k}\Psi_i^{Step2}$ has been chosen for a DOF k by the decision tree of eqn. (5.14), we can continue in the same way as we did for the one-dimensional case described in Chap. 4: A further system of nonlinear equations can be set up, and the solution of this system enables us to parameterize all L trajectory segments ${}^{l}_{k}\vec{m}_i \; \forall \; l \in \{1,\ldots,L\}$ as well as all corresponding time intervals ${}^{l}_{k}\vartheta_i \; \forall \; l \in \{1,\ldots,L\}$ for one selected DOF k. Having done this for every selected DOF, all trajectory parameters of $\mathcal{M}_i$ are finally calculated for the cycle at T_i.

5.1.3 Step 3: Calculation of Output Values

This third step of the OTG algorithm is trivial and works analogously to the calculation of output values in the one-dimensional case. After the complete trajectory $\mathcal{M}_i$ has been calculated for one time step T_i, we only have to find the valid time interval ${}^{\hat{l}}_{k}\vartheta_i$ with $\hat{l} \in \{1,\ldots,L\}$ for each selected DOF $k \in \{1,\ldots,K\}$, such that the condition

$$ {}^{\hat{l}}_{k}t_i \leq \left(T_i + T^{\,cycle}\right) \leq {}^{\hat{l}+1}_{\;\;k}t_i \tag{5.20} $$

is satisfied. The output values $\mathbf{M}_{i+1}$ can finally be calculated by

$$ {}_{k}\vec{M}_{i+1} = {}^{\hat{l}}_{k}\vec{m}_i\left(T_i + T^{\,cycle}\right) \; \forall \; k \in \{1,\ldots,K\} \,, \tag{5.21} $$

such that the final output matrix

$$ \mathbf{M}_{i+1} = \left({}_{1}\vec{M}_{i+1},\ldots,{}_{k}\vec{M}_{i+1},\ldots,{}_{K}\vec{M}_{i+1}\right)^T \tag{5.22} $$

is obtained. If we subsequently call the OTG algorithm again at T_{i+1} and use the same input values, that is, $\mathbf{M}_{i+1}$, $\mathbf{M}_i^{trgt}$, $\mathbf{B}_i$, and $\vec{S}_i$, we will compute exactly the same absolute synchronization time $t_{i+1}^{sync} = t_i^{sync}$, because we compute *the* time-optimal trajectory again. This goes along with the *consistency criterion* of Chap. 3.3 (p. 42). While the absolute value of t_i^{sync} remains constant until the target state of motion has been reached, the time difference from each cycle to the synchronization time naturally decreases by T_{cycle} per cycle:

$$\left(t_i^{sync} - T_i\right) - \left(t_{i+1}^{sync} - T_{i+1}\right) = T_{cycle} \; . \tag{5.23}$$

The resulting values $\mathbf{M}_{i+1}$ are subsequently used for lower-level control.

5.1.4 Final Remarks on the General OTG Algorithm

This section introduced the general OTG algorithm for systems with multiple DOFs, such that all types and variants of OTG (cf. Table 3.1, p. 39) are addressed by this section. The complexity of the algorithm strongly depends on the implemented type: The higher the roman number of the OTG type, the higher is its algorithmic complexity. The measure of complexity is influenced in particular by two properties:

1. The total number
$$2Z+2 = 2\alpha - 4\beta \tag{5.24}$$
of decision trees in Step 1 and Step 2 (cf. Table 3.1, p. 39, and eqn. (5.2), p. 71). Here, in particular the sum of the longest paths — to be more precise: the sum of all computationally most expensive paths — of the decision trees plays the key role.
2. The solvability of the systems of equations, which are generated from the elements in $\mathcal{P}_{Step1}$ and $\mathcal{P}_{Step2}$.

Analogous to the previous chapter, the algorithm for Type IV OTG will be detailed in Sec. 5.3 in order to exemplify the proposed concept by means of a concrete implementation.

5.2 Extension for Variant *B*

The extension from Variant *A* to *B* is very simple. The same Variant *B* decision tree, which has already been introduced for the one-dimensional case, is applied upstream of all $(2Z+2)$ decision trees. Thus, the two cases of eqn. (4.16) and eqn. (4.17) (p. 49) are treated for all selected DOFs $k \in \{1, \ldots, K\}$, and we first bring all DOFs that do not fulfill eqns. (4.16) and (4.17) back into their bounds of $\mathbf{B}_i$. Subsequently the Variant *A* motion profiles transfer all DOFs to their target state of motion $\mathbf{M}_i^{trgt}$.

5.3 Type IV On-Line Trajectory Generation

This section concretizes the proposed general concept of Sec. 5.1 by means of the Type IV on-line trajectory generation algorithm. As we did in Chap. 4.2 (p. 49), we first specify the general parameters of the algorithm based on

Table 3.1 (p. 39): $\alpha = 8$ and $\beta = 3$ indicate that the target acceleration vector $\vec{A}_i^{trgt}$ and the target jerk vector $\vec{J}_i^{trgt}$ equal the zero vector at any time instant T_i with $i \in \mathbb{Z}$ while we can specify a target velocity vector $\vec{V}_i^{trgt}$, which is exactly reached at the target position $\vec{P}_i^{trgt}$. Regarding the motion constraints, maximum velocities $\vec{V}_i^{max}$, maximum accelerations $\vec{A}_i^{max}$, and maximum jerks $\vec{J}_i^{max}$ are taken into account, and each selected DOF k can exhibit up to $Z = 1$ inoperative time interval ${}_k^1\zeta_i$. Four decision trees are required:

- Three decision trees are required for the Step 1 calculations of
  - the minimum execution times ${}_kt_i^{min}$ for every selected DOF $k \in \{1, \ldots, K\}$,
  - the lower limits ${}_kt_i^{begin}$ of the inoperative time interval ${}_k^1\zeta_i$ for every selected DOF $k \in \{1, \ldots, K\}$, and
  - the respective upper limits ${}_kt_i^{end}$ for every selected DOF $k \in \{1, \ldots, K\}$, and
- one decision tree is necessary for Step 2.

We organize this section in the same way as the corresponding Sec. 4.2 of the previous chapter: After the introduction of the Type IV, Variant *A* algorithm, we subsequently describe the algorithm for the *B*-Variant.

A non-scientific problem formulation described in very simple words can again be found in App. D.3 (p. 203).

5.3.1 Type IV, Variant *A*

Step 1

As depicted in the general structogram in Fig. 5.1 (p. 70), it is the goal of this step to calculate the synchronization time t_i^{sync}. After the calculation of the minimum execution times ${}_kt_i^{min}$ for all selected DOFs $k \in \{1, \ldots, K\}$, we subsequently check all selected DOFs to see whether they exhibit an inoperative time interval ${}_k^1\zeta_i$, in order to finally determine the value for t_i^{sync}.

Minimum Execution Time

The procedure for the calculation of the minimum execution times ${}_kt_i^{min}$ for all selected DOFs $k \in \{1, \ldots, K\}$ was already described in Chap. 4.2.1 (p. 50). We apply the decision tree 1*A* of Fig. 4.4 (p. 52) in order to determine an acceleration profile ${}_k\Psi_i^{Step1}$ of the set $\mathcal{P}_{Step1}$. A subset of $\mathcal{P}_{Step1}$ was presented in Fig. 4.3 (p. 50). Once we know the profile ${}_k\Psi_i^{Step1}$, which leads to *the* time-optimal trajectory for DOF k, we can set up a system of nonlinear equations (e.g., eqns. (4.21) – (4.32), p. 53, for the *PosTriNegTri* acceleration profile). Then, a position-error function is generated, which only depends on one of the unknown variables of the system of equations. We are interested in one particular root of this function: The root that delivers *the* time-optimal solution for the given input parameters ${}_k\vec{W}_i$. For this purpose, we determine an interval, from which we know that the root lies within it (e.g., Alg. 4.1, p. 57). After we know the interval bounds, the modified Anderson-Björck-King method (cf. App. A, p. 185) is applied in order to calculate the root numerically. Using this root value, all remaining unknowns of the system of equations can be calculated as exemplarily shown in App. B.3 (p. 190). In contrast to the one-dimensional case, we do not calculate all unknown variables, and we do not parameterize a trajectory, but we only calculate the value of ${}_kt_i^{min}$.

Inoperative Time Intervals

According to Sec. 5.1.1, we have to look for $Z = 1$ inoperative time interval per selected DOF. If such an interval exists for one DOF k, we have to calculate ${}_k^1t_i^{begin}$ and ${}_k^1t_i^{end}$ to determine ${}_k^1\zeta_i$ — the only possible element of ${}_k\mathcal{Z}_i$ (cf. eqn. 5.3). For this purpose, we need two further decision trees, $1B$ and $1C$, one for the calculation of ${}_k^1t_i^{begin}$ and one for ${}_k^1t_i^{end}$. These two trees describe the set of holes

$$\mathcal{H} \subset \mathbb{R}^9 \tag{5.25}$$

in the input domain of the systems of equations for Step 2 (cf. eqn. (5.14)). They work very similarly to the one from Fig. 4.4 (p. 52) and are explained in the following. The result of the first decision tree, $1B$, is an acceleration profile ${}_k\check{\Psi}_i^{Step1} \in \mathcal{P}_{Step1}$ that delivers the system of equations to determine ${}_k^1t_i^{begin}$, the second decision tree, $1C$, selects a profile ${}_k\hat{\Psi}_i^{Step1} \in \mathcal{P}_{Step1}$ in order to compute ${}_k^1t_i^{end}$.

Remark 5.2. *Because these decision trees would engross too much space, they cannot be depicted completely in a book, thesis, or script. To give an impression of the complexity: The tree $1B$ — written in font size of 10 pt, prepared in a minimized version, and with all nodes tightly arranged — can just be plotted on a poster of DIN A0 size.*[1] *Even the description would fill a book of several hundred pages, such that here only an impression shall be imparted, and only the basic conceptual ideas are explained. The cutouts of the four Type IV decision trees presented in Figs. 4.4, 5.6, 5.7, and 5.8 can only be considered as small samples. The decision tree of the extension for Variant B (Fig. 4.8, p. 62), however, is presented in extenso. A deeper discussion on this topic can be found in Chap. 9.10 (p. 173).*

The decision tree $1B$ of Fig. 5.6 determines the acceleration profile for the calculation of ${}_k^1t_i^{begin}$. Compared to the tree $1A$ of Fig. 4.4, in which we tried to reach the target state of motion ${}_k\vec{M}_i^{trgt}$ as fast as possible, we try to reach

[1] The actual $1B$ decision tree as developed by the author engrosses two DIN A0 posters (also in a minimized version and with a font size of 10 pt but with potential for tightening).

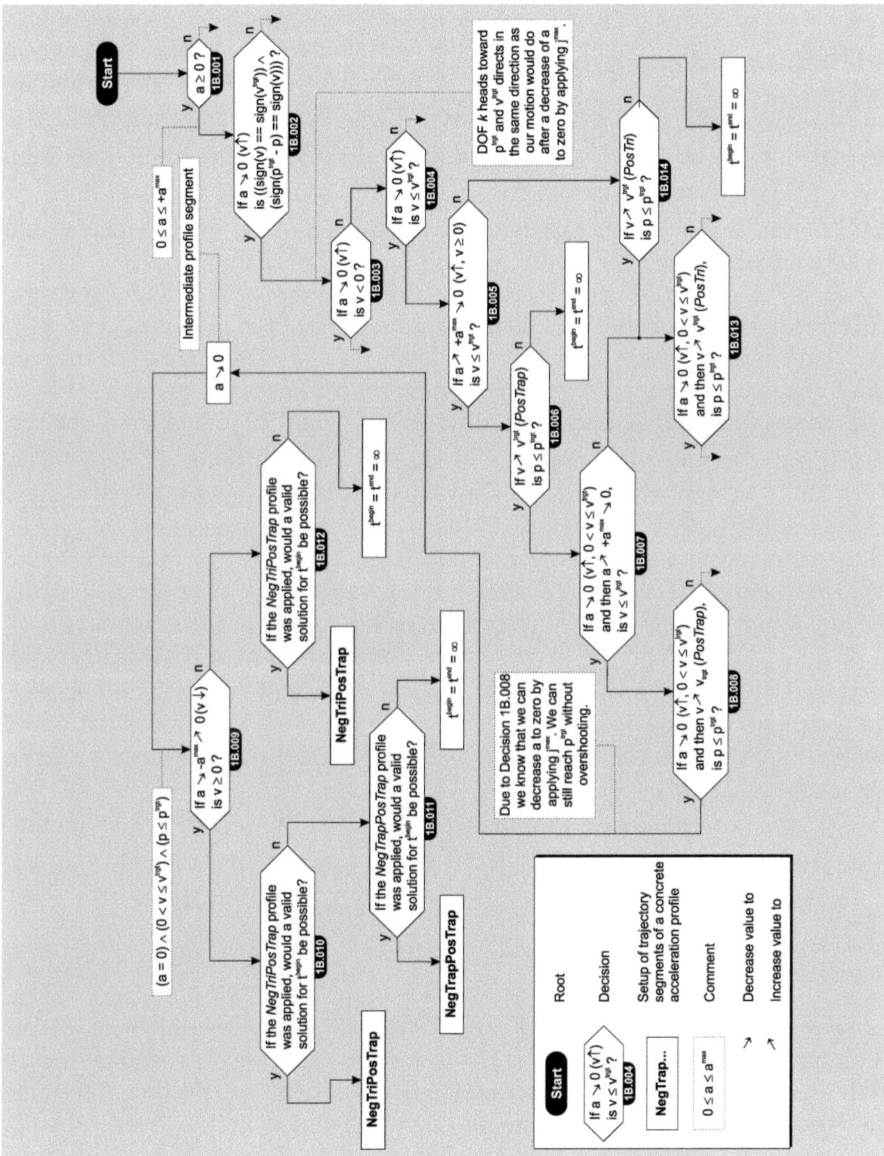

Fig. 5.6 Cutout of the Type IV decision tree $1B$ for the determination of an acceleration profile ${}_k\check{\Psi}_i^{Step1} \in \mathcal{P}_{Step1}$ for the calculation of ${}_kt_i^{begin}$.

${}_k\vec{M}_i^{trgt}$ within the next possible greater time, which can be achieved with another acceleration profile or with another set of profile parameters. The result can either be an acceleration profile ${}_k\check{\Psi}_i^{Step1}$ or the insight that there is no inoperative time interval ${}_k^1\zeta_i$, that is, ${}_k^1t_i^{begin} = \infty$ and ${}_k^1t_i^{end} = \infty$. If the result is an acceleration profile, which we denote with ${}_k\check{\Psi}_i^{Step1}$, we set up the respective system of equations in order to calculate the profile's execution time ${}_k^1t_i^{begin}$. If the system of equations delivers two solutions (cf. Examples 5.1 and 5.2, p. 73), we have to be careful: If the determined profile ${}_k\hat{\Psi}_i^{Step1}$ is the same profile as ${}_k\Psi_i^{Step1}$ that we used to calculate the minimum execution time ${}_k^1t_i^{min}$, we have to take the greater solution; otherwise, that is, ${}_k\hat{\Psi}_i^{Step1} = {}_k\check{\Psi}_i^{Step1}$, the lesser one is the correct one, because the greater one would already have specified the upper interval limit ${}_k^1t_i^{end}$.

Fig. 5.7 depicts the third decision tree for the Type IV OTG algorithm. This tree is only applied if we know from the result of the previously executed tree that an inoperative time interval ${}_k^1\zeta_i$ exists, that is, compared to the tree $1B$ of Fig. 5.6, in tree $1C$ we already know that an upper time interval limit ${}_k^1t_i^{end}$ is present. While tree $1B$ started at ${}_k^1t_i^{min}$ and searched from the left to the right on the time axis to find the left interval limit ${}_k^1t_i^{begin}$, we now start from the right to find ${}_k^1t_i^{end}$. That means we first try to reach the target state of motion ${}_k\vec{M}_i^{trgt}$ of a selected DOF k with a Step 1 acceleration profile, which would require the greatest possible execution time. If the target velocity ${}_kV_i^{trgt}$ is negative (Decision *1C.001*), we try to reach ${}_kP_i^{trgt}$ by first applying the maximum velocity $+{}_kV_i^{max}$ (Decision *1C.004*/*1C.011*). Afterwards, we try to decrease the positive acceleration face *Trap 1C.004* or a *Tri* (*1C.011*) step by step until we find a profile ${}_k\hat{\Psi}_i^{Step1}$ whose system of equations is solvable. If the system of equations delivers two valid solutions for the desired upper interval bound ${}_k^1t_i^{end}$, the maximum one is the correct one, because the lesser one would be ${}_k^1t_i^{begin}$.

Now, we have calculated all minimum execution times ${}_kt_i^{min}$ and all inoperative time intervals ${}_k^1\zeta_i = \left[{}_k^1t_i^{begin}, {}_k^1t_i^{end}\right]$ for all selected DOFs $k \in \{1, \ldots, K\}$, and we know the complete set of all possible candidates for t_i^{sync}.

Determining t_i^{sync}

As shown in the structogram of Fig. 5.1 (p. 70), the determination of the minimum possible value of t_i^{sync} is the last substep of Step 1. It works exactly in the same way as generally described in Sec. 5.1.1 (p. 74). We first determine the greatest element of all minimum execution times, and in a second step, we make sure, that t_i^{sync} is not an element of the set of inoperative time intervals $\mathcal{Z}_i$.

Step 2

In accordance with Sec. 5.1.2, Step 2 calculates the coefficients of all sets of motion polynomials ${}^l\mathbf{m}_i(t)$. That means, we need an acceleration profile ${}_k\Psi_i^{Step2}$ for each single selected DOF k that facilitates the desired time-synchronization, such that all selected DOFs reach their ${}_kP_i^{trgt}$ and ${}_kV_i^{trgt}$

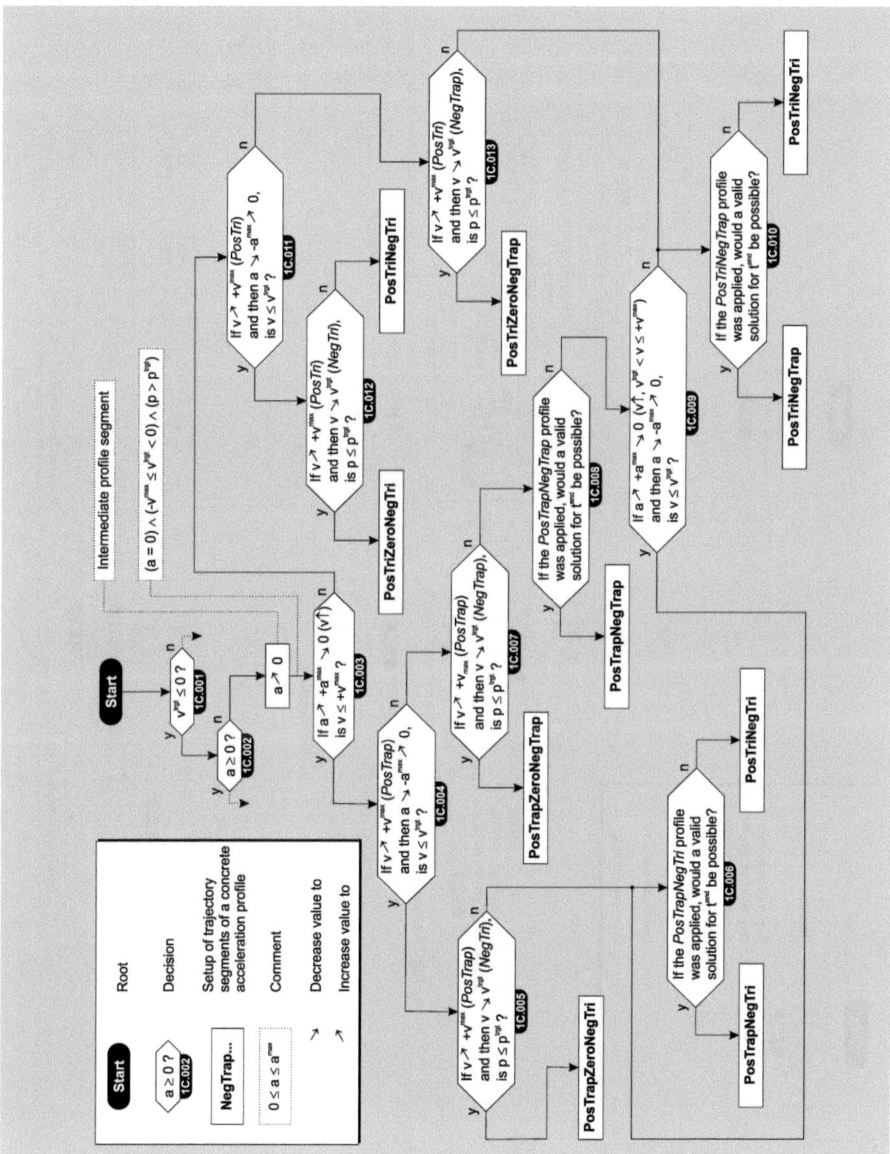

Fig. 5.7 Cutout of the Type IV decision tree 1*C* for the determination of an acceleration profile ${}_{k}\hat{\Psi}_{i}^{Step1} \in \mathcal{P}_{Step1}$ for the calculation of ${}_{k}t_{i}^{end}$.

exactly at t_i^{sync}. Apart from the eight input values of $\mathbf{W}_i$ for each DOF k, t_i^{sync} is the ninth input value for the decision tree from Fig. 5.8, which acts as function

$$f: \left(\mathbb{R}^9 \setminus \mathcal{H}\right) \longrightarrow \mathcal{P}_{Step2}\ , \tag{5.26}$$

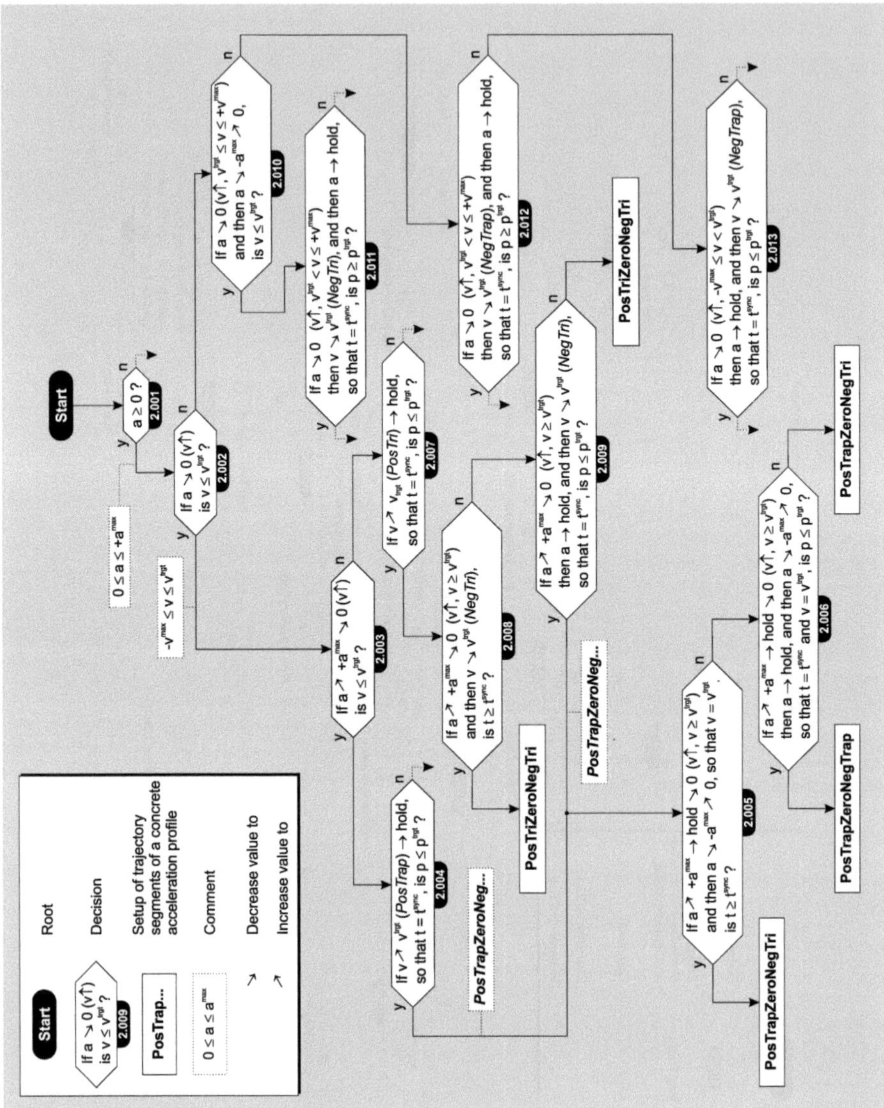

Fig. 5.8 Cutout of the Type IV decision tree for Step 2 to determine an acceleration profile ${}_k\Psi_i^{Step2} \in \mathcal{P}_{Step2}$ for one DOF $k \in \{1, \ldots, K\}$.

that is, the tree has to cover the complete input domain $\mathbb{R}^9 \setminus \mathcal{H}$ for all systems of equations (cf. eqn. (5.26)):

$$^{s}\mathcal{D}_{Step2} \subset \mathbb{R}^9 \tag{5.27}$$

$$\bigcup_{s=1}^{S} {}^{s}\mathcal{D}_{Step2} = \mathbb{R}^9 \setminus \mathcal{H} \,. \tag{5.28}$$

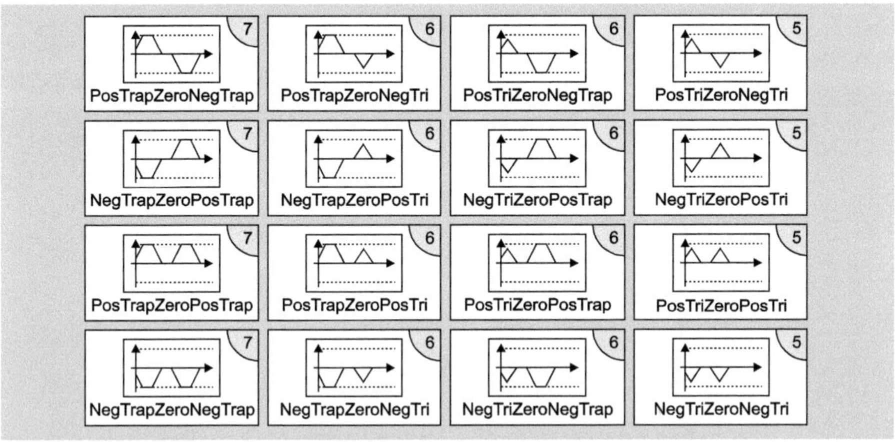

Fig. 5.9 A subset of the acceleration profile set $\mathcal{P}_{Step2}$ of Type IV.

Due to Step 1, we already know that the nine-element input vector is not an element of $\mathcal{H}$. It is important that the optimization criterion of eqn. (5.10) (p. 75) is met by the decision tree and thus by the acceleration profiles of Step 2, $\mathcal{P}_{Step2}$. The essential consequence of eqn. (5.10) is that only rectangular jerk profiles can be applied, and the jerk progression can only take three values: zero, the positive, or the negative maximum jerk value.

A subset of the acceleration profiles of $\mathcal{P}_{Step2}$ is shown in Fig. 5.9, and a selection of these profiles is supposed to be parameterized during Step 2 in order to calculate the trajectory $\mathcal{M}_i$ at a time instant T_i. Fig. 5.8 shows a cutout of the Step 2 decision tree and was drawn analogously to the previous three decision trees of Figs. 4.4, 5.6, and 5.7. As with Decision *1A.001*, Decision *2.001* checks whether the current acceleration value ${}_kA_i$ is positive or negative. Assuming the left branch is selected, we let Decision *2.002* check whether the velocity value would be greater or less than ${}_kV_i^{trgt}$ if $-{}_kJ_i^{max}$ was applied to decrease the acceleration value to zero. Decision *2.003* checks whether ${}_kA_i^{max}$ must be applied to reach ${}_kV_i^{trgt}$, that is, we find out whether a positive triangle or trapezoidal acceleration profile would lead to ${}_kV_i^{trgt}$. We assume that a trapezoidal profile is required. Decision *2.004* inspects, whether the resulting position value of DOF k at t_i^{sync} is greater or less than ${}_kP_i^{trgt}$ if ${}_kV_i^{trgt}$ would be reached as soon as possible by applying a trapezoidal acceleration profile (*PosTrap*). If the resulting value is less, we know that we have to increase the hold time of the trapezoidal acceleration profile, that is, a *PosTrapZeroNeg...* profile would be required. Before we decide, whether ${}_k\Psi_i^{Step2} = PosTrapZeroNegTri$ or ${}_k\Psi_i^{Step2} = PosTrapZeroNegTrap$ is correct, Decision *2.005* verifies whether we would have enough time to apply a positive trapezoidal acceleration profile directly followed by a negative triangle-shaped profile that exactly touches $-{}_kA_i^{max}$, such that ${}_kV_i^{trgt}$ is finally reached. If we do not have enough time for this acceleration progression,

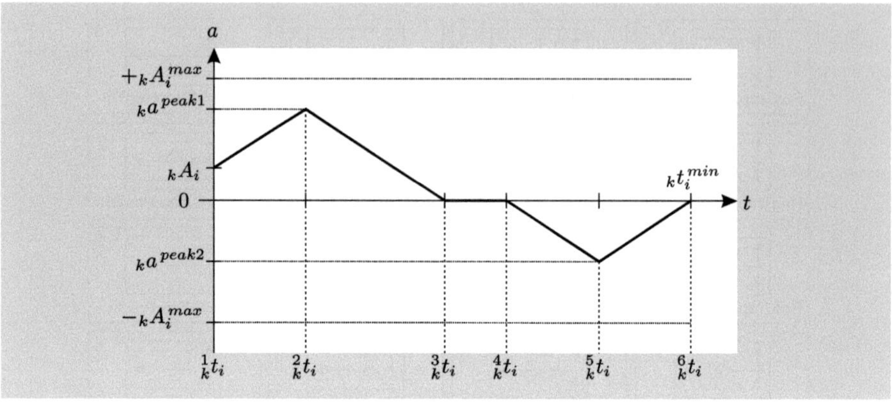

Fig. 5.10 The *PosTriZeroNegTri* profile with all relevant variables, such that the system of equations of eqns. (5.29) – (5.42) can be directly set up.

$_k\Psi_i^{Step2}$ = *PosTrapZeroNegTri* is the solution. If there is enough time for the acceleration behavior checked by Decision *2.005*, Decision *2.006* checks the boundary case for the profiles *PosTrapZeroNegTrap* and *PosTrapZeroNegTri* and finally determines the acceleration profile. The decisions *2.007* to *2.012* work analogously.

Analogous to the system of equations given by eqns. (4.21) – (4.32) (p. 53), we can set up a further system of equations for each acceleration profile $^s\Psi^{Step2} \in \mathcal{P}_{Step2}$ of Fig. 5.9. A Step 2 profile has to be found for every selected DOF $k \in \{1, \ldots, K\}$, and the solutions of all resulting systems of equations deliver all required parameters for all L trajectory segments $^l\mathbf{m}_i \; \forall \; l \in \{1, \ldots, L\}$ as well as all time intervals $^l\mathcal{V}_i$ (cf. eqns. (3.8) and (3.9), p. 34). This procedure will be exemplarily explained by means of the simple profile $^s\Psi^{Step2} = PosTriZeroNegTri$ as shown in Fig. 5.10 (cf. top right element of Fig. 5.9). In a straightforward way we can set up a system of 14 equations for one selected DOF k at time instant T_i:

$$^2_kt_i \; - \; T_i = \frac{\left({_k}a^{peak1} \; - \; {_k}A_i\right)}{{_k}J_i^{max}} \tag{5.29}$$

$$^3_kt_i \; - \; ^2_kt_i = \frac{{_k}a^{peak1}}{{_k}J_i^{max}} \tag{5.30}$$

$$^5_kt_i \; - \; ^4_kt_i = -\frac{{_k}a^{peak2}}{{_k}J_i^{max}} \tag{5.31}$$

$$t_i^{sync} \; - \; ^5_kt_i = -\frac{{_k}a^{peak2}}{{_k}J_i^{max}} \tag{5.32}$$

$$^2_kv_i \; - V_i = \frac{1}{2}\left(^2_kt_i \; - \; T_i\right)\left({_k}A_i + {_k}a^{peak1}\right) \tag{5.33}$$

$$ {}_k^3v_i - {}_k^2v_i = \frac{1}{2}\left({}_k^3t_i - {}_k^2t_i\right)\ {}_ka^{peak1} \tag{5.34} $$

$$ {}_k^4v_i - {}_k^3v_i = 0 \tag{5.35} $$

$$ {}_k^5v_i - {}_k^4v_i = \frac{1}{2}\left({}_k^5t_i - {}_k^4t_i\right)\ {}_ka^{peak2} \tag{5.36} $$

$$ {}_kV_i^{trgt} - {}_k^5v_i = \frac{1}{2}\left(t_i^{sync} - {}_k^5t_i\right)\ {}_ka^{peak2} \tag{5.37} $$

$$ \begin{aligned} {}_k^2p_i - {}_kP_i = {}&{}_kV_i\ \left({}_k^2t_i - T_i\right) + \frac{1}{2}\ {}_kA_i\ \left({}_k^2t_i - T_i\right)^2 \\ &+ \frac{1}{6}\ {}_kJ_i^{max}\ \left({}_k^2t_i - T_i\right)^3 \end{aligned} \tag{5.38} $$

$$ \begin{aligned} {}_k^3p_i - {}_k^2p_i = {}&{}_k^2v_i\ \left({}_k^3t_i - {}_k^2t_i\right) + \frac{1}{2}\ {}_ka_i^{peak1}\ \left({}_k^3t_i - {}_k^2t_i\right)^2 \\ &- \frac{1}{6}\ {}_kJ_i^{max}\ \left({}_k^3t_i - {}_k^2t_i\right)^3 \end{aligned} \tag{5.39} $$

$$ {}_k^4p_i - {}_k^3p_i = {}_k^3v_i\ \left({}_k^4t_i - {}_k^3t_i\right) \tag{5.40} $$

$$ {}_k^5p_i - {}_k^4p_i = {}_k^4v_i\ \left({}_k^5t_i - {}_k^4t_i\right) - \frac{1}{6}\ {}_kJ_i^{max}\ \left({}_k^5t_i - {}_k^4t_i\right)^3 \tag{5.41} $$

$$ \begin{aligned} {}_kP_i^{trgt} - {}_k^5p_i = {}&{}_k^5v_i\ \left(t_i^{sync} - {}_k^5t_i\right) + \frac{1}{2}\ {}_ka_i^{peak2}\ \left(t_i^{sync} - {}_k^5t_i\right)^2 \\ &+ \frac{1}{6}\ {}_kJ_i^{max}\ \left(t_i^{sync} - {}_k^5t_i\right)^3 \end{aligned} \tag{5.42} $$

Equations (5.29) – (5.42) contain 14 unknown variables: ${}_k^2t_i$, ${}_k^3t_i$, ${}_k^4t_i$, ${}_k^5t_i$, ${}_k^2v_i$, ${}_k^3v_i$, ${}_k^4v_i$, ${}_k^5v_i$, ${}_k^2p_i$, ${}_k^3p_i$, ${}_k^4p_i$, ${}_k^5p_i$, ${}_ka_i^{peak1}$, and ${}_ka_i^{peak2}$. The first four values of this list together with T_i and t_i^{sync} compose ${}_k^l\vartheta_i\ \forall\ l\ \in\ \{1,\ldots,5\}$. The latter eight values are used to calculate the motion polynomials at time instant $T_i\ {}_k^l\vec{m}_i\ \forall\ l\ \in\ \{1,\ldots,5\}$, such that the trajectory for DOF k is completely described.

During the solution of these equations, the same numerical problems that we already know of from solving eqns. (4.21) – (4.32) appear. An analytical solution of the systems of equations for the acceleration profiles of $\mathcal{P}_{Step2}$ is again not suitable, such that we apply the same method as described in Chap. 4.2.1 in order to calculate the solution of eqns. (5.29) – (5.42).

First, we manually transform the system of equations to a position-error function

$$ {}^{PosTriZeroNegTri}p_i^{err2}\left({}_ka_i^{peak1}\right) :\ \mathbb{R} \longrightarrow \mathbb{R}\ , \tag{5.43} $$

to which we calculate the root. Depending on the chosen unknown variable, the position-error function ${}^{PosTriZeroNegTri}p_i^{err2}\left({}_ka_i^{peak1}\right)$ is either a monotonically increasing or monotonically decreasing function — hence, we expect only one root. This function is again a transcendental one, and since it is too long, it can be found in its full-length form in App. C.1 (p. 195).

In the same way, we applied Algorithm 4.1 (p. 57) for the function ${}^{PosTriNegTri}p_i^{err}\left(a_i^{peak1}\right)$, we use Algorithm 5.1 to calculate an interval $\left[{}^{min}_{\ k}a_i^{peak1},\ {}^{max}_{\ k}a_i^{peak1}\right]$, from which we know that the function ${}^{PosTriZeroNegTri}p_i^{err2}\left({}_ka_i^{peak1}\right)$ is continuous and contains the only root of the function, that is, the desired value of ${}_ka_i^{peak1}$. It is clear that

$$ {}_kA_i \leq {}^{min}_{\ k}a_i^{peak1} \leq {}_ka_i^{peak1} \leq {}^{max}_{\ k}a_i^{peak1} \leq {}_kA_i^{max} \tag{5.44} $$

holds, but some further restrictions have to be considered and are explained in the following.

In Algorithm 5.1, acceleration profiles are used, such that the synchronization time t_i^{sync} and the boundary values ${}_kA_i^{max}$ and ${}_kJ_i^{max}$ are exerted either as strongly or as weakly as possible. For the first calculation (lines 1 – 4), we set up an acceleration profile, which is described by four segments: $+{}_kJ_i^{max}$ until ${}^{max}_{\ k}a_i^{peak1}$, $-{}_kJ_i^{max}$ until zero, $-{}_kJ_i^{max}$ until ${}^{max}_{\ k}a_i^{peak2}$, and $+{}_kJ_i^{max}$ until zero (line 2). This profile becomes unambiguously specified by the ancillary condition that the profile ends at t_i^{sync} with ${}_kV_i^{trgt}$. After setting up a system of equations for this profile, which we can solve without numerical problems, the two profile parameters ${}^{max}_{\ k}a_i^{peak1}$ and ${}^{max}_{\ k}a_i^{peak2}$ can be calculated in lines 3 and 4. Of course the values are not allowed to exceed the maximum acceleration, such that we simply bound these values in lines 5 – 10. In the next step, we have to find out whether ${}^{max}_{\ k}a_i^{peak1}$ or ${}^{max}_{\ k}a_i^{peak2}$ will be exerted. For this purpose, we calculate the velocity ${}_k\tilde{v}_i$ that we would achieve with the acceleration profile specified in line 12. If ${}_k\tilde{v}_i \leq {}_kV_i^{trgt}$, we know that the positive value ${}^{max}_{\ k}a_i^{peak1}$ of line 3/6 will constitute the upper bound of the acceleration interval, otherwise the value of ${}^{max}_{\ k}a_i^{peak2}$ of line 4/9 has to be considered in the following. In lines 15 – 22 we calculate the lower bound ${}^{min}_{\ k}a_i^{peak1}$; due to line 14 we already know that $\left|{}^{max}_{\ k}a_i^{peak1}\right| \geq \left|{}^{max}_{\ k}a_i^{peak2}\right|$, such that we only have to ensure that ${}_kV_i^{trgt}$ can be reached with the positive acceleration face. If ${}_kV_i^{trgt}$ is exceeded after a decrease of ${}_kA_i$ to zero (line 16), we can simply set ${}^{min}_{\ k}a_i^{peak1}$ to ${}_kA_i$ (line 17). In the other case (lines 19 – 21), we simply calculate the *PosTri* acceleration profile, which would be required to reach ${}_kV_i^{trgt}$. In lines 24 – 27 we already know that $\left|{}^{max}_{\ k}a_i^{peak2}\right| \geq \left|{}^{max}_{\ k}a_i^{peak1}\right| \geq \left|{}^{min}_{\ k}a_i^{peak1}\right|$. Hence, we can set ${}^{min}_{\ k}a_i^{peak1}$ to ${}_kA_i$ (line 24) and calculate ${}^{max}_{\ k}a_i^{peak1}$ in dependence on ${}^{max}_{\ k}a_i^{peak2}$ (line 27) by setting up the profile described in line 26. Due to line 9 and line 14 the resulting upper-bound value ${}^{max}_{\ k}a_i^{peak1}$ cannot be greater than ${}_kA_i^{max}$.

After the execution of Algorithm 5.1, we know that the position-error function of the Step 2 *PosTriZeroNegTri* acceleration profile is continuous within the interval from ${}^{min}_{\ k}a_i^{peak1}$ to ${}^{max}_{\ k}a_i^{peak1}$ and contains exactly one root within it. We use the modified Anderson-Björck-King method of App. A

Algorithm 5.1 Calculate the interval limits for the desired root of ${}_{PosTriZeroNegTri}\,p_i^{err2}\left({}_{k}a_i^{peak1}\right)$.

Require: t_i^{sync}, ${}_{k}V_i^{trgt}$, ${}_{k}V_i$, ${}_{k}A_i^{max}$, ${}_{k}A_i$, ${}_{k}J_i^{max}$, with ${}_{k}A_i^{max} \geq {}_{k}A_i \geq 0$, ${}_{k}J_i^{max} \geq 0$

Ensure: ${}^{min}_{\;\;k}a_i^{peak1}$, ${}^{max}_{\;\;k}a_i^{peak1}$

$\triangleright$ Calculate ${}^{max}_{\;\;k}a_i^{peak1}$ and ${}^{max}_{\;\;k}a_i^{peak2}$ for the profile $\triangleleft$

$\triangleright \left\{{}_{k}A_i \nearrow {}^{max}_{\;\;k}a_i^{peak1} \searrow 0 \searrow {}^{max}_{\;\;k}a_i^{peak2} \nearrow 0,\ \text{so that } t = t_i^{sync} \wedge v = V_i^{trgt}\right\} \triangleleft$

$${}^{max}_{\;\;k}a_i^{peak1} := \frac{3\,({}_{k}A_i)^2 + 2\,{}_{k}A_i\,t_i^{sync}\,{}_{k}J_i^{max} + {}_{k}J_i^{max}\left(\left(t_i^{sync}\right)^2\,{}_{k}J_i^{max} - 4\,{}_{k}V_i + 4\,{}_{k}V_i^{trgt}\right)}{4\,{}_{k}A_i + 4\,t_i^{sync}\,{}_{k}J_i^{max}}$$

$${}^{max}_{\;\;k}a_i^{peak2} := \frac{({}_{k}A_i)^2 - 2\,{}_{k}A_i\,t_i^{sync}\,{}_{k}J_i^{max} + {}_{k}J_i^{max}\left(4\,{}_{k}V_i^{trgt} - \left(t_i^{sync}\right)^2\,{}_{k}J_i^{max} - 4\,{}_{k}V_i\right)}{4\,{}_{k}A_i + 4\,t_i^{sync}\,{}_{k}J_i^{max}}$$

if ${}^{max}_{\;\;k}a_i^{peak1} > {}_{k}A_i^{max}$ **then**
  ${}^{max}_{\;\;k}a_i^{peak1} := {}_{k}A_i^{max}$
end if
if ${}^{max}_{\;\;k}a_i^{peak2} < -{}_{k}A_i^{max}$ **then**
  ${}^{max}_{\;\;k}a_i^{peak2} := -{}_{k}A_i^{max}$
end if

$\triangleright$ Calculate ${}_{k}\tilde{v}_i$ for the profile $\triangleleft$

$\triangleright \left\{{}_{k}A_i \nearrow {}_{k}A_i^{max} \searrow 0 \searrow -{}_{k}A_i^{max} \nearrow 0,\ \text{so that } v = {}_{k}\tilde{v}_i\right\} \triangleleft$

${}_{k}\tilde{v}_i := {}_{k}V_i - \dfrac{({}_{k}A_i)^2}{2\,{}_{k}J_i^{max}}$

if ${}_{k}\tilde{v}_i \leq {}_{k}V_i^{trgt}$ **then**
  $v^{(A_i \searrow 0)} := V_i + \dfrac{(A_i)^2}{2\,J_i^{max}}$
  if $v^{(A_i \searrow 0)} > {}_{k}V_i^{trgt}$ **then**
    ${}^{min}_{\;\;k}a_i^{peak1} := {}_{k}A_i$
  else
    $\triangleright$ Calculate ${}^{min}_{\;\;k}a_i^{peak1}$ for the profile $\triangleleft$
    $\triangleright \left\{{}_{k}A_i \nearrow {}^{min}_{\;\;k}a_i^{peak1} \searrow 0,\ \text{so that } v = V_i^{trgt}\right\} \triangleleft$

$${}^{min}_{\;\;k}a_i^{peak1} := \sqrt{\frac{({}_{k}A_i)^2 + 2\left({}_{k}V_i^{trgt} - {}_{k}V_i\right){}_{k}J_i^{max}}{2}}$$

  end if
else
  ${}^{min}_{\;\;k}a_i^{peak1} := {}_{k}A_i$
  $\triangleright$ Calculate ${}^{max}_{\;\;k}a_i^{peak1}$ for the profile $\triangleleft$
  $\triangleright \left\{{}_{k}A_i \nearrow {}^{max}_{\;\;k}a_i^{peak1} \searrow 0 \searrow {}^{max}_{\;\;k}a_i^{peak2} \nearrow 0,\ \text{so that } v = V_i^{trgt}\right\} \triangleleft$

$${}^{max}_{\;\;k}a_i^{peak1} := \sqrt{\frac{({}_{k}A_i)^2 + 2\left({}^{max}_{\;\;k}a_i^{peak2}\right)^2 + 2\left({}_{k}V_i^{trgt} - {}_{k}V_i\right){}_{k}J_i^{max}}{2}}$$

end if

(p. 185) to calculate the desired value of ${}_{k}a_{i}^{peak1}$. The practical relevance of the Type III OTG algorithm of Haschke et al. [108] suffers from incorrect solutions caused by numerical inaccuracies. Due to the procedure proposed here, the algorithm becomes robust against numerical instabilities, which is indispensable for a reliable usage.

Thus, we obtain the first of the 14 unknowns of eqns. (5.29) – (5.42). The calculations of the other 13 variables are trivial and can be found in App. C.2 (p. 196). App. C.2 furthermore describes the determination of all trajectory parameters for one selected DOF $k \in \{1, \ldots, K\}$:

$$\forall\, l \in \{1, \ldots, 5\}:$$

$$ {}_{k}^{l}\vec{m}_i(t) = \left({}_{k}^{l}p_i(t), {}_{k}^{l}v_i(t), {}_{k}^{l}a_i(t), {}_{k}^{l}j_i(t)\right) \tag{5.45}$$

$$ {}_{k}^{l}\vartheta_i \in {}^{l}\mathcal{V}_i \qquad \text{with } {}_{k}^{l}\vartheta_i = \left[{}_{k}^{l}t_i, {}^{l+1}_{\;\;k}t_i\right] . \tag{5.46}$$

Through eqns. (5.45) and (5.46), we parameterize all values of $\mathcal{M}_i$, which regard DOF k, that is, ${}_{k}\mathcal{M}_i$ (cf. eqns. (3.9) and (3.10), p. 34).

If this is done for all selected DOFs as shown in the structogram of Fig. 5.1 (p. 70), the trajectory at time instant T_i, $\mathcal{M}_i$, is completely described.

Step 3

The last step of the Type IV OTG algorithm works exactly as described in Sec. 5.1.3 (p. 79), and we can calculate the output values $\vec{P}_{i+1}$, $\vec{V}_{i+1}$, and $\vec{A}_{i+1}$ for the current control cycle. These values are subsequently used for lower-level control.

Example of Parameterizing the Step 2 *PosTriZeroNegTri* Acceleration Profile

To clarify the previously introduced concepts for on-line trajectory generation for systems with multiple DOFs, we apply a concrete example again and explain all required steps for one translational DOF. We take the *PosTriNeg-Tri* example of Chap. 4.2.1 (p. 59) as a base and use the same input values again (cf. eqn. (4.41)):

$$\left.\begin{array}{l}
\left.\begin{array}{rcl}
{}_{1}P_0 &=& -499\,mm\\
{}_{1}V_0 &=& -335\,mm/s\\
{}_{1}A_0 &=& 152\,mm/s^2
\end{array}\right\}\ {}_{1}\vec{M}_0\\
\left.\begin{array}{rcl}
{}_{1}P_0^{trgt} &=& -90\,mm\\
{}_{1}V_0^{trgt} &=& -347\,mm/s
\end{array}\right\}\ {}_{1}\vec{M}_0^{trgt}\\
\left.\begin{array}{rcl}
{}_{1}V_0^{max} &=& 985\,mm/s\\
{}_{1}A_0^{max} &=& 972\,mm/s^2\\
{}_{1}J_0^{max} &=& 324\,mm/s^3
\end{array}\right\}\ {}_{1}\vec{B}_0\\
\begin{array}{rcl}
{}_{1}S_0 &=& 1
\end{array}
\end{array}\right\}\ {}_{1}\vec{W}_0 . \tag{5.47}$$

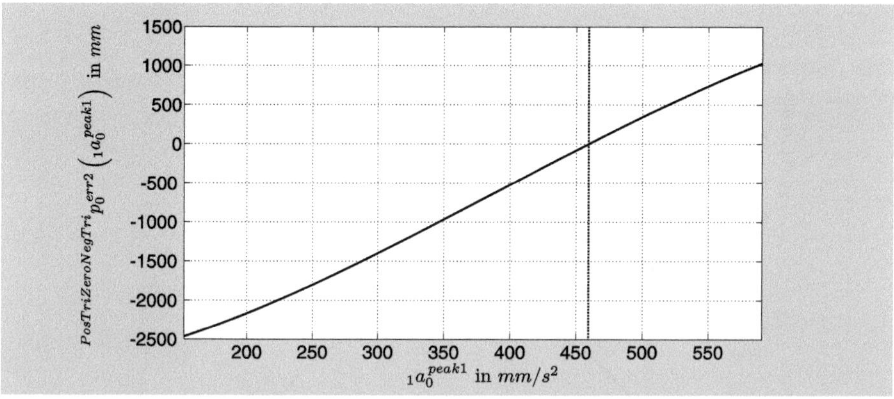

Fig. 5.11 Progression of the Step 2 position-error function ${}^{PosTriZeroNegTri}p_0^{err2}\left({}_1a_0^{peak1}\right)$ for the input values of eqns. (5.47) and (5.49) at T_0 within the interval $[152.000\,mm/s^2, 591.651\,mm/s^2]$.

In the first step of the algorithm, we have to calculate the minimum execution time ${}_1t_i^{min}$, and as we already know from eqn. (4.45), a minimum execution time of

$$ {}_1t_i^{min} = 5.795\,s \tag{5.48}$$

will be required if ${}_1t_i^{min}$ determines t_i^{sync}. In this case, Step 2 would calculate exactly the same trajectory as depicted in Fig. 4.7 (p. 61) for DOF 1. To explain the functionality of the synchronization concept of Step 2, we assume that another selected DOF requires a greater synchronization time t_i^{sync}. Let the (arbitrarily chosen) value be

$$ t_i^{sync} = 6.795\,s\ , \tag{5.49}$$

that is, one second more, such that eqns. (5.47) and (5.49) constitute the input parameters for the Step 2 decision tree of Fig. 5.8. The resulting acceleration profile is the *PosTriZeroNegTri* profile, that is, we can set up eqns. (5.29) – (5.42) in order to calculate the desired trajectory parameters. After the transformation to the position-error function ${}^{PosTriZeroNegTri}p_i^{err2}\left({}_1a_i^{peak1}\right)$, Algorithm 5.1 is executed to compute the interval, in which the desired value of ${}_1a_i^{peak1}$ can be found:

$$ {}^{min}_{\ \ 1}a_0^{peak1} = 152.000\,mm/s^2 \quad \text{and} \quad {}^{max}_{\ \ 1}a_0^{peak1} = 591.651\,mm/s^2\ . \tag{5.50}$$

This interval is used to apply the modified Anderson-Björck-King method of App. A (p. 185) to calculate the root of the position-error function. In the illustration of Fig. 5.11, one can clearly see that the function monotonically increases within the calculated interval. The resulting root

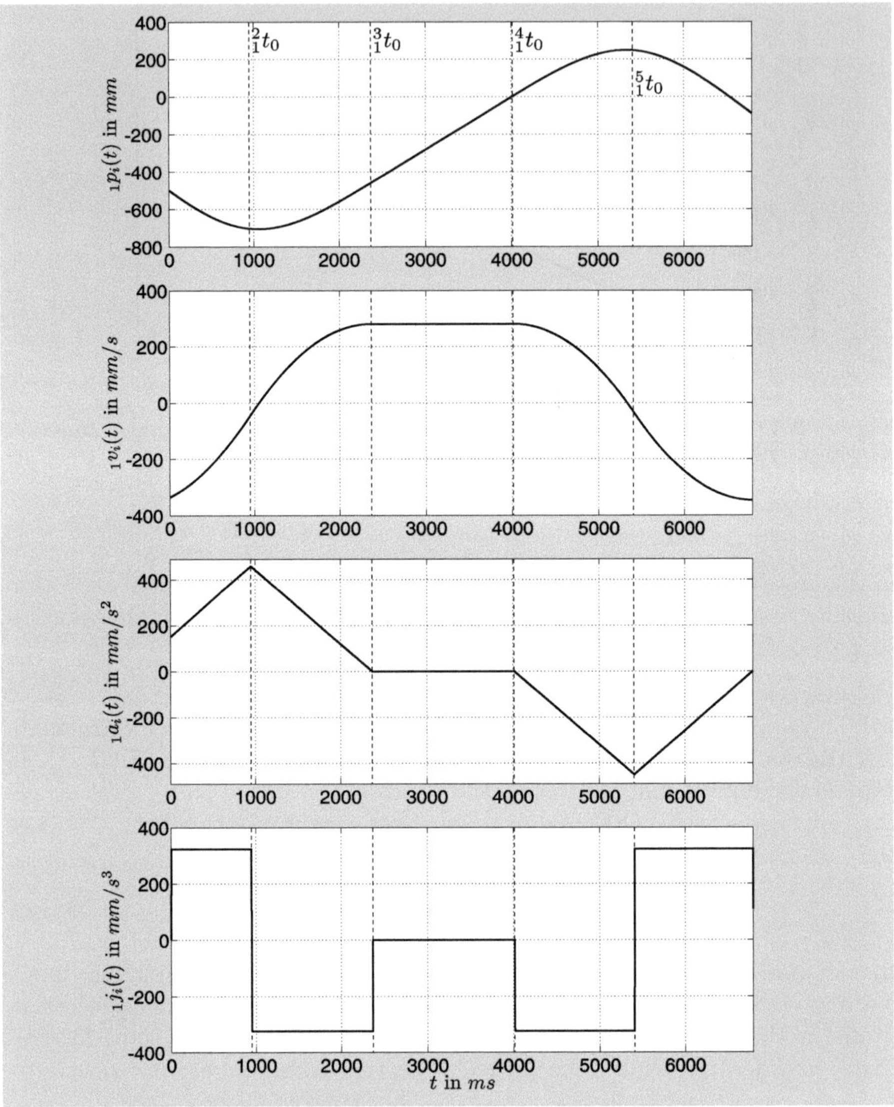

Fig. 5.12 Resulting trajectory for DOF 1 with $i \in \{0, \ldots, 6795\}$ for the given input values of eqns. (5.47) and (5.49). The dashed lines indicate the times $^2_1t_0$, $^3_1t_0$, $^4_1t_0$, and $^5_1t_0$ calculated in eqn. (5.52).

$$_1a_0^{peak1} = 459.259\,mm/s^2 \tag{5.51}$$

determines the first of the 14 unknowns (cf. eqn. (4.44) and Fig. 4.6). As in Step 1, we subsequently can compute the remaining 13 unknown variables as described in App. C.2 (p. 196, eqns. (C.15) – (C.27)):

$$
\begin{array}{llll}
{}_1^2t_0 = 0.948\,s & & {}_1^3t_0 = 2.366\,s & \\
{}_1^4t_0 = 4.012\,s & & {}_1^5t_0 = 5.403\,s & \\
{}_1^2v_0 = -45.163\,mm/s & & {}_1^3v_0 = 280.328\,mm/s & \\
{}_1^4v_0 = 280.328\,mm/s & & {}_1^5v_0 = -33.336\,mm/s & \quad (5.52) \\
{}_1^2p_0 = -702.287\,mm & & {}_1^3p_0 = -458.722\,mm & \\
{}_1^4p_0 = 2.771\,mm & & {}_1^5p_0 = 247.356\,mm & \\
{}_1a_0^{peak2} = -450.838\,mm/s^2\ . & & &
\end{array}
$$

As a last substep the final trajectory parameters $\mathcal{M}_i$ for DOF 1 are calculated by applying eqns. (C.28) – (C.52). Thus, we have calculated all polynomial coefficients ${}_k^l\vec{m}_i\ \forall\ l\ \in\ \{1,\ldots,5\}$ as well as all respective time intervals ${}_1^l\vartheta_i$ with $l\ \in\ \{1,\ldots,5\}$. When we plot all these polynomials for the time instants from $T_0\ =\ 0\,ms$ to $t_i^{sync}\ =\ 6795\,ms$, we obtain the diagrams of Fig. 5.12.

5.3.2 Type IV, Variant B

As we already know from Sec. 5.2, the extension from Variant A to Variant B is very simple. The Variant B decision tree of Fig. 4.8 (p. 62) is applied upstream of all Variant A trees. Thus, all decision trees and respective systems of equations can cope with varying elements in $\mathbf{B}_i$ as described for the Type IV, Variant B OTG algorithm in Sec. 4.2.2 (p. 63). Up to Λ trajectory segments are selected and parameterized for each selected DOF. These trajectory segments are executed prior to the Variant A acceleration profiles. Thus, we can predict that each DOF will be brought back into its constraints in the shortest possible time, and that afterwards the Variant A acceleration profiles will ensure that all selected DOFs reach $\mathbf{M}_i^{trgt}$ synchronously.

5.4 Summary and Final Remarks

We introduced the main OTG algorithm for systems with multiple degrees of freedom in Variant A and in Variant B. Subsequent to the general description of the three required algorithmic steps in Sec. 5.1, the algorithm of Type IV, Variant A was exemplarily presented in order to explain necessary details in a more concrete way. The B-Variants fulfill all four criteria, which were introduced in Chap. 3.3 (p. 40). As these four criteria are of fundamental character for the OTG algorithm, we briefly summarize the interrelationship of this chapter to the four criteria of Chap. 3.3:

(i) **Criterion for Time-Optimality.** After the minimum possible execution times and all inoperative time intervals have been calculated for all selected DOFs, the minimum possible synchronization time is chosen in Step 1, that is, there is no possibility of reaching the target state of motion $\mathbf{M}_i^{trgt}$ before this time t_i^{sync}.

This criterion is fulfilled due to Step 1.

(ii) Criterion for Time-Synchronization. Step 2 of the algorithm ensures that all selected DOFs reach their target state of motion exactly at the synchronization time.

This criterion is fulfilled due to Step 2.

(iii) Criterion for Motion Constraints. For the case where current motion values exceeded their respective bounds, or where it is unavoidable that they will exceed their bounds at future time instants, Variant B trajectory segments are applied to guide all motion values back into their bounds. Afterwards, the Variant A motion profiles ensure that these bounds are not exceeded again.

This criterion is fulfilled due to the separation of Variant A and Variant B.

(iv) Criterion for Consistency. The fulfillment of this criterion must be split into two items:

1. We always calculate *the* time-optimal trajectory. If we calculate a trajectory at a time instant T_i, Step 1 attains a certain synchronization time t_i^{sync} for *the* time-optimal trajectory. If the algorithm is executed with the same input values, that is, $\mathbf{M}_{i+1}$, $\mathbf{M}_i^{trgt}$, $\mathbf{B}_i$, and $\vec{S}_i$, again at T_{i+1} we will obtain exactly the same synchronization time $t_{i+1}^{sync} = t_i^{sync}$, because we compute *the* time-optimal trajectory again (cf. eqn. (5.23), p. 80).

   This part of the criterion is fulfilled due to Step 1.
2. In the one-dimensional case, it is clear that the first item is sufficient, because there can always be only one solution for the trajectory with the minimum execution time t_i^{min}. But in the multi-dimensional case, all DOFs $k \in \{1, \ldots, K\}$, which have to finish their motion at a time instant greater than ${}_kt_i^{min}$, could principally be transferred to ${}_k\vec{M}_i^{trgt}$ by an infinite number of possible trajectories. To cope with this problem, a certain trajectory criterion is taken into account to solve an additional optimization problem. Thus, the synchronization procedure of Step 2 works unambiguously, that is, for any given set of input parameters, exactly one trajectory can be calculated.

   This part of the criterion is fulfilled due to the optimization functions of eqns. (5.9) – (5.11) (p. 75).

To point out the three most important scientific insights of this chapter:

Inoperative Time Intervals

Depending on the type of OTG algorithm, each selected DOF $k \in \{1, \ldots, K\}$ can have up to $Z = \alpha - 2\beta - 1$ inoperative time intervals, in which its target state of motion ${}_k\vec{M}_i^{trgt}$ cannot be reached (cf. eqn. (5.2), p. 71).

Equations (5.18) and (5.19)

$$\bigcup_{s=1}^{S} {}^{s}\mathcal{D}_{Step2} = \left(\mathbb{R}^{\alpha+1}\right) \setminus \mathcal{H}$$

$$\mathcal{S} = \bigcup_{s=1}^{S} \left\{ {}^{s}\mathcal{D}_{Step2} \cap {}^{u}\mathcal{D}_{Step2} \; \forall \; u \in \{1, \ldots, S\} \middle| u \neq s \right\}$$

Inoperative time intervals lead to holes $\mathcal{H}$ in the input domain space of the trajectory synchronization algorithm (eqn. (5.18)). These holes $\mathcal{H}$, as well as the input domains of the systems of equations for the motion profiles of Step 2, are bounded by α-dimensional hyperplanes $\mathcal{S}$ in the $(\alpha+1)$-dimensional space (eqn. (5.19)).

Decision Trees

The OTG algorithm requires $2Z+2 = 2\alpha - 4\beta$ decision trees (cf. eqn. (5.24)): one tree to calculate the minimum execution time of each selected DOF $k \in \{1, \ldots, K\}$, $2Z$ trees to calculate the up to Z inoperative time intervals and therewith also $\mathcal{H}$, and one tree to calculate the synchronized trajectories. All trees together describe $\mathcal{S}$ (cf. eqn. (5.19)).

The algorithmic concept of command variable generation presented here can be applied in a wide range of robotic applications. Furthermore, new architectural concepts in the field of hybrid switched-system control become possible. Both aspects are discussed in Chap. 7. Examples and interesting results are shown in Chap. 8. But before, the next chapter introduces a very relevant special case for on-line trajectory generators: straight-line trajectories.

Chapter 6
On-Line Generation of Homothetic Trajectories

The proposed algorithm of the previous chapter constitutes the heart of this book and will be slightly extended in this chapter. Here, we consider a special case that is a very relevant one in practice: on-line generation of homothetic trajectories. These trajectories are one-dimensional straight-lines in a multi-dimensional space and are relevant for all straight-line motion operations in robotics. For the realization of this additional feature, the trajectories generated by the OTG algorithm not only have to be time-synchronized, but also *phase-synchronized*. After the problem has been formulated in the following section, an adaptation of the general OTG algorithm that copes with this new demand is presented in Sec. 6.2.

6.1 Problem Formulation

As we know from Chap. 5, there is principally an infinite number of possible trajectories that can fulfill the criterion of time-synchronization. Due to the simple optimization functions of eqns. (5.9) – (5.11) (p. 75), we were able to develop a unique and consistent concept for time-synchronization (Step 2). But the usage of the proposed optimization functions is — of course — not the only possibility. We can also apply other optimization criteria, and a particular method that lets the algorithm generate straight-line (homothetic) trajectories is defined in the following.

The textbook of Kostrikin, Manin, and Alferieff [134] gives a good introduction to homothety, and Khalil and Dombre [126] apply it to robot trajectory generation. To generate homothetic trajectories in K-dimensional space, we can take an arbitrary DOF $\kappa \in \{1,\ldots,K\}$ as reference DOF and design the trajectory parameters, such that the condition

$$\forall\ (k,l) \in \{1,\ldots,K\} \times \{1,\ldots,L\} : \quad {}_{k}^{l}v_i(t) = {}_{k}\rho_i\ {}_{\kappa}^{l}v_i(t) \text{ with } t \in {}_{k}^{l}\vartheta_i \tag{6.1}$$

is fulfilled. This obviously implies that also

T. Kröger: On-Line Trajectory Generation in Robotic Systems, STAR 58, pp. 99–104.
springerlink.com

$$\forall\ (k,l)\ \in\ \{1,\ldots,K\}\times\{1,\ldots,L\}:\qquad \left.\begin{array}{l} {}_{k}^{l}a_i(t) = {}_{k}\rho_i\ {}_{\kappa}^{l}a_i(t) \\ {}_{k}^{l}j_i(t) = {}_{k}\rho_i\ {}_{\kappa}^{l}j_i(t) \\ {}_{k}^{l}d_i(t) = {}_{k}\rho_i\ {}_{\kappa}^{l}d_i(t) \end{array}\right\}\ \text{with}\ t\ \in\ {}_{k}^{l}\vartheta_i \tag{6.2}$$

are fulfilled. The ratios between the reference DOF κ and all other DOFs $\{1,\ldots,K\}\backslash\{\kappa\}$ are defined by the constant vector

$$\vec{\rho}_i = \left({}_{1}\rho_i,\ldots,{}_{k}\rho_i,\ldots,{}_{K}\rho_i\right)^T \quad \text{with} \quad {}_{\kappa}\rho_i = 1\ . \tag{6.3}$$

Usually, homothetic trajectories are generated as described by eqns. (6.1) to (6.3): A scalar function specifies the velocity progression for one DOF, which is referred to as reference DOF, and the motions of the other DOFs are calculated by using $\vec{\rho}_i$ [126].

The proposed OTG algorithm of the previous chapter does not allow the generation of homothetic trajectories, which are very important. Fig. 6.1 illustrates a simple two-DOF point-to-point motion with zero-velocities in the initial and in the target position:

$$\vec{P}_0 = \begin{pmatrix}50\\50\end{pmatrix} mm \qquad \text{and} \qquad \vec{P}_0^{trgt} = \begin{pmatrix}800\\300\end{pmatrix} mm\ . \tag{6.4}$$

One can clearly recognize the differences between the phase-synchronized (homothetic), the time-synchronized (cf. Chap. 5), and the non-synchronized trajectory. The elements of the boundary values $\mathbf{B}_i$

$$\vec{V}_0^{max} = \begin{pmatrix}80\\80\end{pmatrix} mm/s \quad \vec{A}_0^{max} = \begin{pmatrix}250\\250\end{pmatrix} mm/s^2 \quad \vec{J}_0^{max} = \begin{pmatrix}400\\400\end{pmatrix} mm/s^3 \tag{6.5}$$

are the same for both DOFs, such that the *non-synchronized* motion starts with an angle of 45° with respect to the reference frame. The *time-synchronized* one starts in $\vec{P}_0$ *and* ends in $\vec{P}_0^{trgt}$ with an angle of 45°. To generate

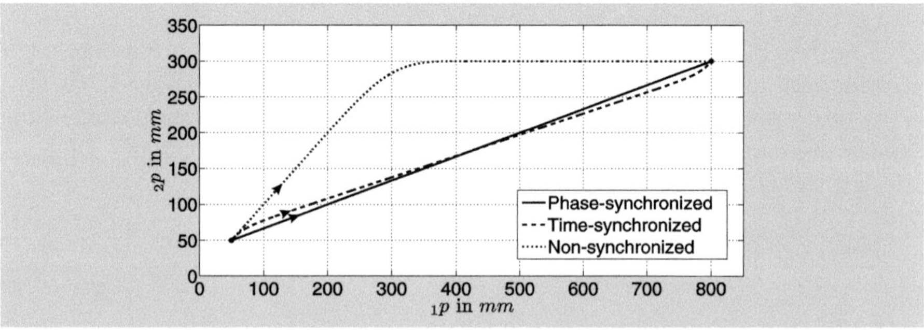

Fig. 6.1 XY-plots of a plain two-DOF point-to-point motion: phase-synchronized, time-synchronized, and without synchronization.

phase-synchronized (homothetic) trajectories on-line, we have to calculate the internal parameter vector $\vec{\rho}_i$ (eqns. (6.1) – (6.3)) and adapt the boundary motion parameters $\mathbf{B}_i$ to a matrix $\mathbf{B}'_i$. Whether and how the given boundary parameters $\mathbf{B}_i$ can be adapted to $\mathbf{B}'_i$ in order to achieve a straight-line trajectory in a multi-dimensional space is explained in the following section.

6.2 The Algorithm

Here, we derive the extension of the existing general OTG algorithm, whose input and output values were shown in Fig. 3.3 (p. 37). It is the task of the OTG algorithm to time-optimally transfer some current state of motion $\mathbf{M}_i$ into the target state of motion $\mathbf{M}_i^{trgt}$ taking into consideration boundary values $\mathbf{B}_i$. In contrast to this general case, the homothetic trajectories cannot be generated from arbitrary states of motion. The basic requirement for homothety is that the vectors $\vec{V}_i$, $\vec{A}_i$, $\vec{J}_i$, $\vec{V}_i^{trgt}$, $\vec{A}_i^{trgt}$, $\vec{J}_i^{trgt}$, and $\left(\vec{P}_i^{trgt} - \vec{P}_i\right)$ are collinear[1]:

Condition 1

$\exists\, \vec{\gamma} = \left(\gamma_1, \ldots, \gamma_{(\alpha-\beta-2)}\right) \in \mathbb{R}^{(\alpha-\beta-2)}$, that is, for the Type IX OTG algorithm:

$$\left(\vec{P}_i^{trgt} - \vec{P}_i\right) = \gamma_1\, \vec{V}_i = \gamma_2\, \vec{V}_i^{trgt} = \gamma_3\, \vec{A}_i = \gamma_4\, \vec{A}_i^{trgt} = \gamma_5\, \vec{J}_i = \gamma_6\, \vec{J}_i^{trgt}\ . \quad (6.6)$$

In correspondence to Table 3.1 (p. 39), $1 \leq \alpha - \beta - 2 \leq 6$ holds. If the input values comply with the condition of eqn. (6.6), we have to select a reference DOF κ, which we use in the following to adapt the respective elements of $\mathbf{B}_i$. To sustain the feature of (kinetic) time-optimality, we first calculate the execution times of all selected DOFs

$$\vec{t}_i^{min} = \left({}_1t_i^{min}, \ldots, {}_kt_i^{min}, \ldots, {}_Kt_i^{min}\right)^T \quad (6.7)$$

and the respective motion profiles

$$\vec{\Psi}_i^{Step1} = \left({}_1\Psi_i^{Step1}, \ldots, {}_k\Psi_i^{Step1}, \ldots, {}_K\Psi_i^{Step1}\right)^T \quad (6.8)$$

$$\text{with}\quad {}_k\Psi_i^{Step1} \in \mathcal{P}_{Step1} \quad \forall\, k \in \{1, \ldots, K\}\ .$$

Fig. 6.2 shows the general Nassi-Shneiderman structogram of the algorithm presented here. Based on eqn. (6.7), we can determine the maximum element of $\vec{t}_i^{min}$, which we define as ${}_\kappa t_i^{min}$.

$$_\kappa t_i^{min} = \max\left\{{}_1t_i^{min}, \ldots, {}_Kt_i^{min}\right\} \quad (6.9)$$

[1] For the OTG Types I-II, $\vec{A}_i$, $\vec{A}_i^{trgt}$, $\vec{J}_i$, and $\vec{J}_i^{trgt}$ are irrelevant; for the Types III-V, $\vec{J}_i$ and $\vec{J}_i^{trgt}$ are not relevant, because they are not considered by these OTG algorithms (cf. Table 3.1, p. 39).

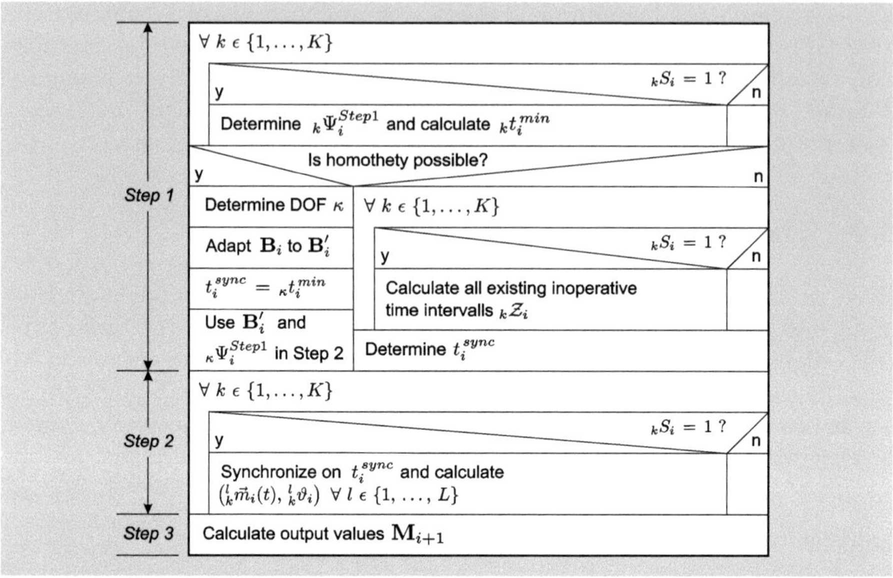

Fig. 6.2 General Nassi-Shneiderman structogram for the OTG algorithm for homothetic trajectories (cf. Fig. 5.1, p. 70).

If phase-synchronization is possible, κ will become the reference DOF for the current trajectory (cf. eqns. (6.1) – (6.3)), and the synchronization time t_i^{sync} will be set to ${}_{\kappa}t_i^{min}$.

$$t_i^{sync} = {}_{\kappa}t_i^{min} \tag{6.10}$$

Apart from the first requirement of eqn. (6.6), the single vectors of $\mathbf{B}_i$ also have to be collinear to the vectors of eqn. (6.6):

$$\exists\ \vec{\delta} = \left(\delta_1, \ldots, \delta_\beta\right) \in \mathbb{R}^\beta\ , \tag{6.11}$$

that is, for the Types VI – IX of the OTG algorithm ($\beta = 4$)

$\forall\ k \in \{1, \ldots, K\}:$

$$\left|{}_kP_i^{trgt} - {}_kP_i\right| = \delta_1\ {}_kV_i^{max} = \delta_2\ {}_kA_i^{max} = \delta_3\ {}_kJ_i^{max} = \delta_4\ {}_kD_i^{max}\ . \tag{6.12}$$

Since this is commonly not given, we have to adapt $\mathbf{B}_i$ to $\mathbf{B}_i'$, or more precisely: all elements of $\mathbf{B}_i$ except the ones of the reference DOF κ. This is done in two steps. First, we calculate the vector

$$\vec{\rho}_i = \frac{\vec{P}_i^{trgt} - \vec{P}_i}{\left|{}_{\kappa}P_i^{trgt} - {}_{\kappa}P_i\right|}\ , \tag{6.13}$$

whose κ-th element ${}_{\kappa}\rho_i$ is one. Subsequently, the adapted elements of $\mathbf{B}_i'$ can be calculated by

$$\forall\ k\ \in\ \{1,\ldots,K\}:$$

$$\begin{aligned} {}_{k}V_i^{max\,\prime} &= {}_{k}\rho_i\ {}_{\kappa}V_i^{max} & {}_{k}J_i^{max\,\prime} &= {}_{k}\rho_i\ {}_{\kappa}J_i^{max} \\ {}_{k}A_i^{max\,\prime} &= {}_{k}\rho_i\ {}_{\kappa}A_i^{max} & {}_{k}D_i^{max\,\prime} &= {}_{k}\rho_i\ {}_{\kappa}D_i^{max}. \end{aligned} \tag{6.14}$$

With

$$\mathbf{B}_i' = \left(\vec{V}_i^{max\,\prime},\ \vec{A}_i^{max\,\prime},\ \vec{J}_i^{max\,\prime},\ \vec{D}_i^{max\,\prime}\right) \tag{6.15}$$

the requirement of eqn. (6.12) is fulfilled. In order not to breach the boundary values $\mathbf{B}_i$ the elements of $\mathbf{B}_i'$ must be less than or equal the original ones of $\mathbf{B}_i$:

Condition 2

$$\forall\ k\ \in\ \{1,\ldots,K\}:$$

$$\begin{aligned} & {}_{k}V_i^{max\,\prime} \leq {}_{k}V_i^{max} & \wedge \quad & {}_{k}J_i^{max\,\prime} \leq {}_{k}J_i^{max} \\ \wedge \quad & {}_{k}A_i^{max\,\prime} \leq {}_{k}A_i^{max} & \wedge \quad & {}_{k}D_i^{max\,\prime} \leq {}_{k}D_i^{max}\ . \end{aligned} \tag{6.16}$$

If this condition is also fulfilled, one further condition has to be satisfied to be able to generate a homothetic trajectory. All selected DOFs have to be executed with the motion profile of the reference DOF ${}_{\kappa}\Psi_i^{Step1}$ to comply with eqns. (6.1) and (6.2). As we know from Chap. 4 and Chap. 5, we can use ${}_{\kappa}\Psi_i^{Step1}$ to set up a system of equations for each selected DOF. But instead of $\mathbf{B}_i$, $\mathbf{B}_i'$ becomes applied here together with $\mathbf{M}_i$, $\mathbf{M}_i^{trgt}$, and $\vec{S}_i$. The solution of a system of equations contains the execution time ${}_{k}t_i^{min\,\prime}$ for a selected DOF k. For all DOFs except κ, the value differs from ${}_{k}t_i^{min}$

$${}_{k}t_i^{min\,\prime} \geq {}_{k}t_i^{min} \qquad \forall\ k \in \{1,\ldots,K\}\backslash\{\kappa\} \tag{6.17}$$

$${}_{\kappa}t_i^{min\,\prime} = {}_{\kappa}t_i^{min}\ , \tag{6.18}$$

because $\mathbf{B}_i'$ and ${}_{\kappa}\Psi_i^{Step1}$ are applied. It may happen, that no valid solution can be found for one or more DOFs, because the system of equations for the profile ${}_{\kappa}\Psi_i^{Step1}$ does deliver a valid solution. In such a case, the profile ${}_{\kappa}\Psi_i^{Step1}$ cannot transfer $\mathbf{M}_i$ to $\mathbf{M}_i^{trgt}$ and homothety is not possible. To generate a homothetic trajectory, all selected DOFs have to reach their target state of motion synchronously, that is,

Condition 3

$${}_{k}t_i^{min\,\prime} = {}_{\kappa}t_i^{min} \qquad \forall\ k\ \in\ \{1,\ldots,K\} \tag{6.19}$$

must be fulfilled as a final condition. Equation (6.19) has to be regarded as theoretical. Due to numerical inaccuracies

$$\left| {}_k t_i^{min\,\prime} - {}_\kappa t_i^{min} \right| \leq T^{cycle} \qquad \forall\; k \in \{1,\ldots,K\} \tag{6.20}$$

has to be applied in practice. T^{cycle} is the cycle time, that is, the time interval, in which the OTG algorithm is periodically executed.

As a result of the above, the **Conditions 1**, **2**, and **3** (eqns. (6.6), (6.16), and (6.20)) must be fulfilled to generate a homothetic trajectory. In Fig. 6.2, this check is done in the block *"Is homothety possible?,"* that is, if possible, the OTG generates a homothetic trajectory (left branch), otherwise only a time-synchronized trajectory is generated (right branch). In the case of homothety, the profile ${}_\kappa\Psi_i^{Step1}$ and the adapted boundary values $\mathbf{B}_i'$ are applied in Step 2 to all selected DOFs in order to calculate homothetic motion parameters for the trajectory $\mathcal{M}_i(t)$ at time instant T_i.

This chapter presented an adaptation of the general OTG algorithm of Chap. 5 in order to provide the possibility of generating trajectories along a one-dimensional straight-line in a multi-dimensional space during runtime, that is, the possibility to *instantaneously react to unforeseen and abrupt set-point switchings* is sustained also for this kind of trajectory. This adapted algorithm can be applied on-line in the same way as the OTG algorithm of the previous chapter, and it fulfills all four criteria (i – iv) of Chap. 3.3 (p. 40). Concrete experimental results are presented in Chap. 8.3 (p. 141).

Chapter 7
Hybrid Switched-System Control for Robotic Systems

Following the concept of on-line trajectory generation introduced in the last three chapters, let us now apply this concept to a robot control system. The goal of this chapter is to present a hybrid switched-control system, which enables the integration of multiple sensors and — in particular — the execution of sensor-guided *and* sensor-guarded robot motion commands.

7.1 Hybrid Switched-System Control

Dynamical systems that are described by an interaction between continuous and discrete dynamics are usually called *hybrid systems* [163]. In Chap. 2.4.3 (p. 24), a very brief overview of hybrid switched-systems was given. Fig. 2.3 (p. 24) illustrates such a hybrid switched-system for a one-DOF system. This figure is extended by Fig. 7.1, such that it becomes suitable for application in robot motion control. Fig. 7.1 was designed for the control of robotic manipulator arms, as they are common for industrial use. This type of serial kinematic machine is chosen to demonstrate the application of the OTG algorithm, because the author's experimental environment is based on such a robot system, and because we can address the majority of robots currently running on the planet. But there is no question that the same ideas can be applied to other kinematic structures or even to systems with only one DOF. Even if very advanced systems, as the DLR light-weight arms [8, 112], which feature torque feedback signals for impedance control [149], are considered, the same basic idea can be applied with only slight changes.

We assume a standard hardware setup of standard industrial manipulators: The robotic system at the bottom left corner is equipped with position sensors (encoders or resolvers) and is connected to a set of servo drive controllers, one per DOF, whose control structure is depicted in a simplified manner [160]. The third element of Fig. 7.1 is a real-time computing system with a respective interface to servo drive controllers, for example, *Sercos III*™ [231], *EtherCAT*™ [74], or even an analog interface with motion control board. The

T. Kröger: On-Line Trajectory Generation in Robotic Systems, STAR 58, pp. 105–135.
springerlink.com

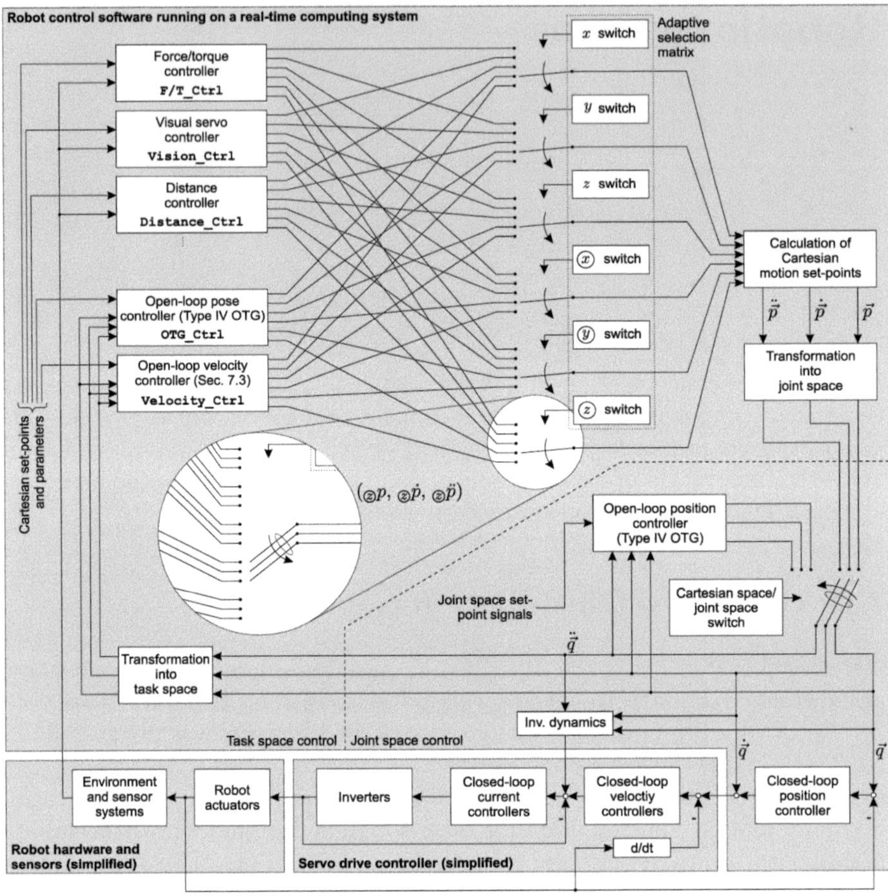

Fig. 7.1 Schematic robot control architecture based on a hybrid switched-system, in which the OTG algorithm of Chap. 5 and Chap. 6 acts as one submodule. Furthermore, it is applied for command variable generation in actuator space.

joint space control concept is also a very classical one. Details on these kinds of control schemes can be found in abundance in the literature [55, 127, 135, 246, 278].

All blocks of the top part of the figure are software components, and the OTG Type IV algorithm is used twice and provides two different functionalities in this hybrid switched-system:

OTG as open-loop pose control submodule in a switched-system
The block *Open-loop pose controller (Type IV OTG)* applies the Type IV OTG algorithm and has the same input and output values as depicted in Fig. 3.3 (p. 37). In this example, we assume the robot to be equipped with a force/torque sensor, a vision system, and a distance sensor. Corresponding to these sensor systems, we use three closed-loop controllers for

sensor-guided motion control in the task space. A transfer function is assigned to each closed-loop controller, and all DOFs that are *not* controlled by one of the closed-loop controllers are guided by the OTG algorithm. The block *Open-loop velocity controller* will be explained later in Sec. 7.3. As a consequence, we require a certain supervisory switching unit that selects the respective controllers for each DOF adapted to the currently executed task and guarantees stability. This is performed by the *adaptive selection matrix*, which will be introduced as part of the Manipulation Primitive Framework in the next section. After a controller has been chosen for each DOF, a new state of motion in Cartesian space $(\vec{p}, \dot{\vec{p}}, \ddot{\vec{p}})$ with $\vec{p} = ({}_{x}p, {}_{y}p, {}_{z}p, {}_{\textcircled{x}}p, {}_{\textcircled{y}}p, {}_{\textcircled{z}}p)^T$ can be calculated. $({}_{x}p, {}_{y}p, {}_{z}p)$ represent the position, and $({}_{\textcircled{x}}p, {}_{\textcircled{y}}p, {}_{\textcircled{z}}p)$ represent the orientation in Cartesian space. After its transformation into joint space, the values of $(\vec{q}, \dot{\vec{q}}, \ddot{\vec{q}})$ act as command variables for lower-level control.

This kind of task level control is the basic structure that enables the execution of sensor-guided *and* sensor-guarded robot motion; we are now able to switch the control method of each DOF at any arbitrary (e.g., sensor-dependent) time instant. In particular: We are able to switch from sensor-guided robot motion control to trajectory-following control at any time, such that the execution of sensor-guarded motion becomes possible (cf. Def. 1.1, p. 5). Another new feature is the possibility of arbitrary reference frame changes: We are able to change the pose of the reference frame from one control cycle to the next one.

OTG as command variable generator in actuator (joint) space

The second instance of OTG in Fig. 7.1 is dedicated to the control of serial kinematic machines. Parallel or hybrid kinematic machines would require a different architecture. This instance of the OTG algorithm enables switchings from task space control to joint space control at any arbitrarily chosen time instant and in any arbitrary state of motion. This important feature cannot be provided without the concept of OTG. Some exemplary advantages of this idea are:

- Possibility of a desired or undesired instantaneous aborting of a task space motion due to an unforeseen sensor event (e.g., sensor near metering range limit, human being in workspace, or arbitrarily changed task parameters) and safe continuation in joint space.
- Instantaneous and safe aborting of a (sensor-guided) task space motion due to sensor malfunctions.
- Possibility of aborting a high-performance motion in task space at sensor-dependent system states and immediate continuation in joint space (neither a stop nor a continuous path motion is required).
- For the case that a sensor-guided motion is heading toward a singularity, the Cartesian motion can be aborted and safely continued in joint space.

These switchings, either in Cartesian space from one controller to another or the switching from Cartesian space control to joint space control (or vice versa), play a very fundamental role for task specification, programming, and system stability. To discuss this in the following, we briefly describe the manipulation primitive framework.

Remark 7.1. *Both OTG instances can be considered as a special case of the classic transition window technique, which is commonly used for continuous path motions [98, 168, 205, 222, 256]. The particularity of the OTG algorithm is that the transition window between two motion segments has a size of* zero.

Remark 7.2. *In earlier works [85, 88, 140, 141, 142], the two described switching features were realized with a Type I OTG algorithm, which is only qualified for experimental works. The Type I properties are comparable to driving a motorcar either at full throttle, at full braking, or without touching the brake or the gas pedal. The resulting rectangular acceleration signals lead to high wear of all mechanical components and excite natural vibrations, such that the lifetimes of robotic systems (and also of cars) would be significantly decreased. Due to the algorithms presented in this book, in particular the Type IV OTG algorithm, systems as shown in Fig. 7.1 become ready for practice.*

7.2 The Manipulation Primitive Framework

The robot motion control and robot motion specification framework descried in the following was proposed by Finkemeyer [85] in 2004. The content of this section is based on his ideas, approaches, and experiments, and the author pays great tribute to him here. In order to demonstrate the potential of multi-sensor integration and the execution of sensor-guided and sensor-guarded robot motion commands — but in particular for the conclusiveness of this monogrph — we briefly repeat some parts of this framework in order to explain how the switchings in Fig. 7.1 work in detail.[1]

The concept of Manipulation Primitives is based on the works of Mason [176], De Schutter et al. [227, 228], and Bruyninckx [39, 40]. As shown in the overview of Chap. 2.4.3 (p. 23), many research groups have acted on these works. Concrete works on the regarded concept were published by Mosemann and Wahl [187, 188], who introduced the concept of Skill Primitives for robot assembly tasks. Finkemeyer and the author of this book [88, 137, 138, 139] focused on control aspects in this context and introduced the term *Manipulation Primitive* (MP). Milighetti and Kuntze [182, 183, 184] extended the MP concept with further elements that enable the control system to make a certain fuzzy- and/or probability-based switching decision during task execution. Thomas et al. [259, 261] applied the MP concept to robot assembly tasks, and Maaß et al. [172, 173] as well as Reisinger [220] transferred it to parallel kinematic machines.

[1] Please note that a different notation is used in this chapter. Details and an overview of all used symbols can be found in the list of symbols in the preface.

The following two subsections introduce MPs and the necessary details on the underlying hybrid switched-system control scheme. Compared to [85] the control scheme becomes modified here. While the original MP concept only works pose-/position-based, which leads to non-optimal trajectory-following behavior and also non-optimal dynamic properties, we now apply the same algorithms in state space.

7.2.1 Manipulation Primitives as Interface to Hybrid Switched-Systems

This subsection introduces MPs as an interface to hybrid switched-systems and gives a formal definition, which will be important for the next subsection, in which the concept of unambiguous mapping of MP parameters to a hybrid switched-system control scheme, as shown in Fig. 7.1, will be explained.

An MP at time instant T_i is formally defined as the three-tuple

$$\mathcal{MP}_i := \{\mathcal{HM}_i, \tau_i, \lambda_i\} \tag{7.1}$$

where $\mathcal{HM}_i$ defines a hybrid motion, τ_i contains tool commands, and the stop condition λ_i determines the end of a single MP. The tool command τ_i is not relevant for this work[2]; the other two quantities will be described and discussed in the following.

Hybrid Move $\mathcal{HM}_i$

$\mathcal{HM}_i$ defines a hybrid move in the sense of the Task Frame Formalism [40]. In the classical case, a motion command is given w.r.t. the Task Frame and is determined by a six-dimensional vector representing the *Compliance Frame* [176]. This simple vector is not a sufficient — and in particular — not a practical way to parameterize sensor-guided robot motion commands when considering multiple sensors.

Example 7.1. *A transition from free space into any contact state requires a switch from one controller, for example, a feedforward trajectory-following controller, to an adequate and task-dependent one, for example, a force/torque controller. This switching can only be realized on the low-level control layer (cf. [31] and [163]), that is, in* real-time *from one control cycle to the next one — instantaneously at the moment of contact detection. Of course, the same behavior must be achieved in the opposite case, that is, a switch from force/torque control to trajectory-following control and/or the case of contact loss. Hence, the responsibility of discretely switching in-between a set of continuously working discrete controllers cannot be shifted to the user or the user application but must necessarily be executed on this low control level.*

[2] Details can be found in [85] and in [88].

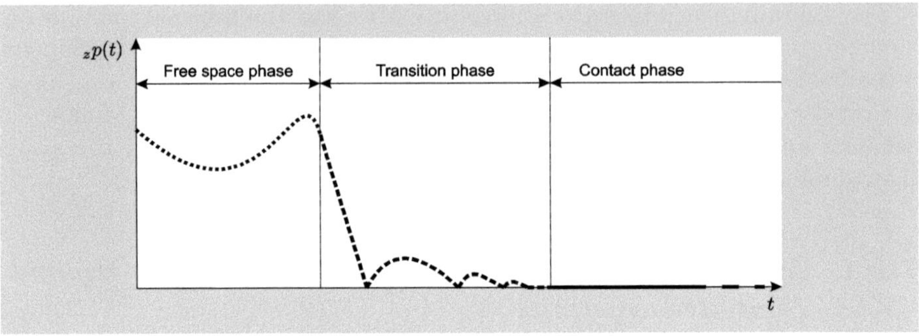

Fig. 7.2 Three phases of establishing an environmental contact [37, 220] (cf. Example 7.1).

The hybrid move command $\mathcal{HM}_i$ is an interface to freely specify the control behavior on this low level.

Furthermore, it is easily realizable to utilize contact transition controllers *as proposed by Brogliato [37] and Reisinger [220]. An intuitive example on this issue is shown in Fig. 7.2.*

The question of how to specify motion commands unambiguously and universally, such that any kind and any number of sensors can be addressed, is supposed to be answered in this context. For this purpose, we define the hybrid motion command $\mathcal{HM}_i$ at instant T_i as

$$\mathcal{HM}_i := \{\mathcal{TF}_i, \mathcal{D}_i\} \tag{7.2}$$

with

$$\mathcal{TF}_i := \{\vec{\theta}_i, RF_i, ANC_i, FFC_i\} \tag{7.3}$$

$$\vec{\theta}_i = ({}_x\theta_i, {}_y\theta_i, {}_z\theta_i, {}_{\text{ⓧ}}\theta_i, {}_{\text{ⓨ}}\theta_i, {}_{\text{ⓩ}}\theta_i)^T \in \mathbb{R}^6 \tag{7.4}$$

$$RF_i, ANC_i \in \{HF, WF, BF, EF\} \tag{7.5}$$

$$FFC_i \in \{WF, BF, EF\} \tag{7.6}$$

and

$$\mathcal{D}_i := \left\{ {}^l_kD^c_i = \left\{ {}^l_k\Psi^c_i, {}^l_k\Phi^c_i \right\} \mid (k, l, c) \in (\mathcal{K} \times \mathcal{L} \times \mathcal{C}) \right.$$
$$\left. \wedge \, \forall \, c \in \mathcal{C} \, \exists \, (w, j) \in \mathcal{K} \times \mathcal{L} \right\} \tag{7.7}$$

with

$$\mathcal{K} := \{x, y, z, \circledx, \circledy, \circledz\} \tag{7.8}$$

$$\mathcal{L} := \{0, \ldots, (m-1)\} \tag{7.9}$$

$$\mathcal{C} := \{0, \ldots, (m-1)\}$$
$$= \{\texttt{OTG\_Ctrl}, \texttt{Velocity\_Ctrl}, \texttt{F/T\_Ctrl}, \texttt{Vision\_Ctrl}, \ldots\} \,. \tag{7.10}$$

Equations (7.2) – (7.10) are discussed in the following. The hybrid move command $\mathcal{HM}_i$ of eqn. (7.2) is specified w.r.t. the Task Frame $\mathcal{TF}_i$, in which a set of set-points $\mathcal{D}_i$ is applied. The following part explains the Task Frame definition (eqns. (7.3) – (7.6)), and subsequently details on $\mathcal{D}_i$ are introduced (eqns. (7.7) – (7.10)).

Definition of the Task Frame w.r.t. Different Reference and Anchor Frames

Fig. 7.3 shows all respective frame assignments for the context of this concept. At each instant T_i with $i \in \mathbb{Z}$, a state of motion is assigned to each frame

$${}^{A}\mathbf{M}_i^{B} = \left({}^{A}\vec{P}_i^{B}, {}^{A}\vec{V}_i^{B}, {}^{A}\vec{A}_i^{B}\right) , \tag{7.11}$$

that is, the motion of frame B w.r.t. frame A. The work cell is assumed to be equipped with n different sensors, of which $\tilde{n}$ are stationary mounted in the robot's environment WF (world frame), $(\hat{n}-\tilde{n})$ are mounted w.r.t. the robot hand HF, and $(n-\hat{n})$ are mounted on some external system EF[3]. If sensor systems are *stationary* mounted w.r.t. some system, their respective motion states only contain a pose value and two zero vectors. If the robot is stationary mounted in its work cell, the velocity and acceleration vector of ${}^{WF}\mathbf{M}_i^{BF}$ are also equivalent to the zero vector for all $i \in \mathbb{Z}$. The motion states of all frames are updated every control cycle, and *it is the aim of the hybrid switched-system in Fig. 7.1 to calculate a new motion state of the hand frame HF w.r.t. the robot base frame BF:* ${}^{BF}\mathbf{M}_i^{HF}$. This motion state can be subsequently transformed into joint space by using the inverse kinematic equation and the inverse or pseudo-inverse Jacobian.

The Task Frame at instant T_i, $\mathcal{TF}_i$, is specified by eqn. (7.3) and can be anchored by the ANC_i frame either to the robot hand frame HF, to the world frame WF, or to the external frame EF. During the execution of one single MP the transformation ${}^{ANC}\vec{P}_i^{TF}$ remains constant. The vector $\vec{\theta}_i$ (eqn. (7.3)) determines the position $({}_{x}\theta_i, {}_{y}\theta_i, {}_{z}\theta_i)$ and orientation $({}_{\circledx}\theta_i, {}_{\circledy}\theta_i, {}_{\circledz}\theta_i)$ of the Task Frame w.r.t. the reference frame RF_i at the *beginning of a single MP execution.* RF_i can be selected from a set of frames, that is known by the system (eqn. (7.5)). Pose, velocity, and acceleration of each of these frames

[3] Of course it is possible to use more than one external system; in this example, we only introduce one system EF in order to explain its handling.

is updated every control cycle T^{cycle}. The frame ANC_i serves as anchor and connects the Task Frame rigidly to another frame of the work cell. The frame FFC_i (eqns. (7.3) and (7.6)) usually equals the world frame or the robot base frame if a mobile manipulation system is considered. If (compliant) motions are to be executed w.r.t. an external moving coordinate system, or if the manipulator is mounted on a mobile platform, the frame FFC_i (i.e., its pose, velocity, and acceleration) enables the internal computation of feedforward compensation (FFC) signals, such that a sensor-based motion command can be executed in dynamic systems in the same way as in static ones (cf. [261]). Regarding the anchor frame, we distinguish between different cases, which are briefly discussed in the following.

Task Frame Attached to Hand Frame ($ANC_i = HF$)
This case is comparable to the classic Compliance Frame Formalism [40, 176, 227] and hybrid force/pose control [217]. Here the Task Frame is synonymous to the *Center of Compliance*; it follows the motions of the hand frame of the manipulator. All set-points must be interpreted *relatively* in order to allow unequivocal hybrid switched-system control. Hence, the position and orientation set-points do not determine the target Task Frame pose w.r.t. the old Task Frame, they determine the relative displacements of the Task Frame along the corresponding DOFs. If rotations are involved, the final pose depends on angular velocities, angular accelerations, and

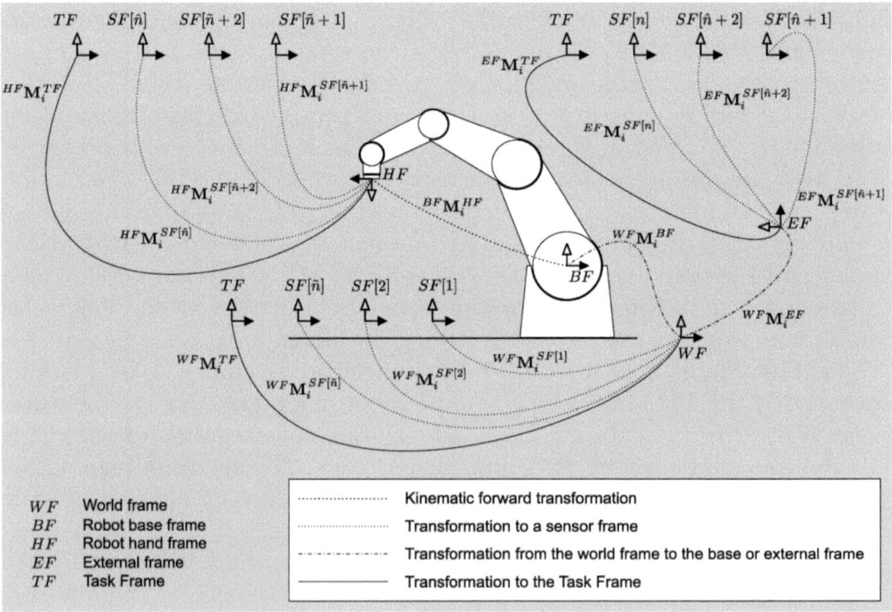

Fig. 7.3 Frame assignments in a robot work cell in accordance with [88] for the proposed method of motion specification.

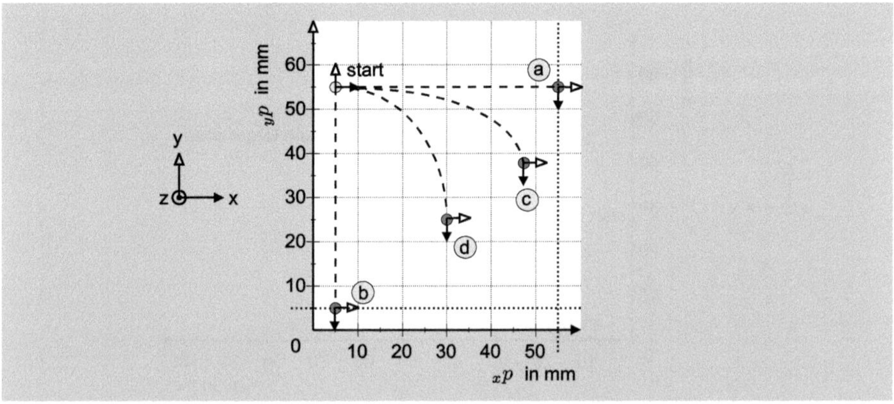

Fig. 7.4 Task Frame attached to the hand frame ($ANC_i = HF$): Depending on maximum and/or desired angular velocities ${}_{ⓩ}V_i^{max}$ and accelerations ${}_{ⓩ}A_i^{max}$ (jerks are assumed to be infinite in this simple example, that is, Type I OTG), the pose set-points ${}_xP_i^{trgt} = 50mm$, ${}_yP_i^{trgt} = 0mm$, and ${}_{ⓩ}P_i^{trgt} = 90°$ result in different Task Frame poses depicted from a) to d): Only the covered distance is the same in all cases; a) ${}_{ⓩ}A_i^{max} \longrightarrow 0°/s^2$, ${}_{ⓩ}V_i^{max} \longrightarrow 0°/s$, ${}_xA_i^{max} \longrightarrow \infty$, and ${}_xV_i^{max} \longrightarrow \infty$; b) ${}_{ⓩ}A_i^{max} \longrightarrow \infty$, ${}_{ⓩ}V_i^{max} \longrightarrow \infty$, ${}_xA_i^{max} \longrightarrow 0mm/s^2$, and ${}_xV_i^{max} \longrightarrow 0mm/s$; c) somewhere between a) and b); d) same as c) (cf. [88]).

angular jerks. For a better understanding, Fig. 7.4 shows a simple example with three DOFs.

Task Frame Attached to Robot Base Frame ($ANC_i = BF$)
The Task Frame pose w.r.t. the robot base frame is fixed during the execution of MPs. In this configuration, two different options are available:

(a) Pose set-points are interpreted in an *absolute* way and determine the target pose of the hand frame. Classical robot move commands can be realized with this kind of Task Frame configuration. Hybrid switched-system control is not possible.
(b) Pose set-points are interpreted *relatively*, and hybrid switched-system control becomes enabled. The difference between both modes is illustrated by the simple example in Fig. 7.5. Of course it is possible to switch between both modes at any time instant.

Task Frame Attached to World Frame ($ANC_i = WF$)
If the manipulator's base is stationary mounted within its work cell, the control behavior is the same as in the previous case. In the case of mobile manipulation systems, the motion is executed w.r.t. the stationary Task Frame. Hybrid switched-system control is possible in the same way as it is for the $ANC_i = BF$ case.

Task Frame Attached to External Frame ($ANC_i = EF$)
The external frame EF indicates a frame of any device (e.g. a conveyor belt, another manipulator, or a mobile robot). This is similar to the previous

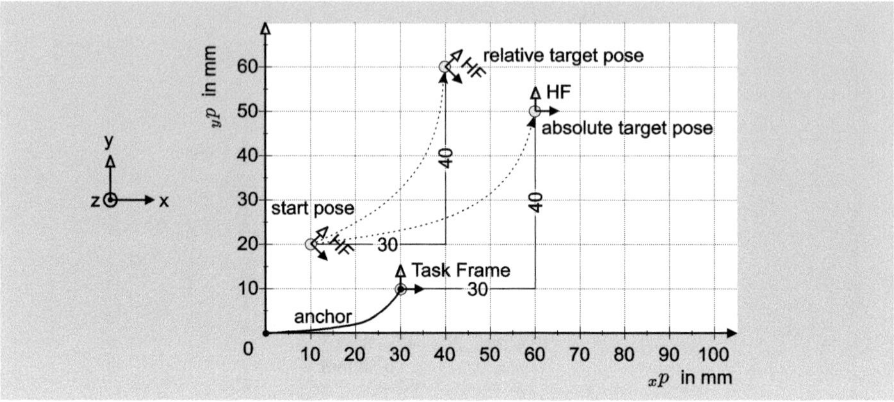

Fig. 7.5 Task Frame attached to the world frame ($ANC_i = WF$): The hand frame reaches different poses if the pose set-points ${}_xP_i^{trgt} = 30mm$, ${}_yP_i^{trgt} = 40mm$, and ${}_{ⓩ}P_i^{trgt} = 0°$ are interpreted either *relatively* or *absolutely* (cf. [88]).

case. However, the anchor frame ANC_i is not mandatorily fixed, that is, the Task Frame may follow a moving frame EF.

Definition of the Set-Point Set $\mathcal{D}_i$

Let us now discuss eqns. (7.7) – (7.10). The set $\mathcal{D}_i$ (eqn. (7.7)) contains up to $(6m)$ set-points ${}_k^lD_i^c$, where m is the number of available controllers. Each controller of the set $\mathcal{C}$ (eqn. (7.10)) is identified by the index c; $\mathcal{C}$ contains $|\mathcal{C}| = m$ controllers. Since it is possible to use more controllers than sensors,

$$n \leq m \tag{7.12}$$

holds. A set-point ${}_k^lD_i^c$ for DOF k (eqn. (7.8)) on control level l (eqn. (7.9)) that is assigned to controller c (eqn. (7.10)) consists of two attributes:

- a set of set-points ${}_k^l\Psi_i^c$ and
- a set of (optional) parameters ${}_k^l\Phi_i^c$.

Remark 7.3. *For the case of* $c =$ `OTG_Ctrl`*: The sets* ${}_k^l\Psi_i^{\texttt{OTG\_Ctrl}}\ \forall\ k \in \left\{x, y, z, ⓧ, ⓨ, ⓩ\right\}$ *would contain the elements of the desired target state of motion* $\mathbf{M}_i^{trgt}$*, and the parameters* ${}_k^l\Phi_i^{\texttt{OTG\_Ctrl}}\ \forall\ k \in \left\{x, y, z, ⓧ, ⓨ, ⓩ\right\}$ *would contain the elements of the boundary values* $\mathbf{B}_i$ *(cf. Chaps. 3 – 6).*

Equation (7.7) assures that only one controller can be assigned for every combination of DOF k and level l. Table 7.1 presents an example of a set-point set $\mathcal{D}_i$. It demonstrates that a list of alternative devices can be declared for each DOF. Of course, these alternatives are optional and its number depends on the current task. They are unambiguously indicated by the level

Table 7.1 Tabular example of the MP set-point set $\mathcal{D}_i$. The physical units depend on the selected controller. For each DOF, $(m-1)$ alternatives can be specified to ensure deterministic and stable behavior (cf. [85]).

level l	${}^{l}_{x}\Psi^{c}_{i}$	controller c	${}^{l}_{y}\Psi^{c}_{i}$	controller c	${}^{l}_{z}\Psi^{c}_{i}$	controller c
0	30	`OTG_Ctrl` in mm	-15	`F/T_Ctrl` in N	20	`Distance_Ctrl` in mm
1	—	—	0.0	`Distance_Ctrl` in mm	15.0	`Velocity_Ctrl` in mm/s
2	—	—	30	`Vision_Ctrl` in mm	—	—
...	...	...	...	...	...	...
$m-1$	—	—	—	—	—	—

level l	${}^{l}_{ⓧ}\Psi^{c}_{i}$	controller c	${}^{l}_{ⓨ}\Psi^{c}_{i}$	controller c	${}^{l}_{ⓩ}\Psi^{c}_{i}$	controller c
0	10	`Vision_Ctrl` in °	1	`F/T_Ctrl` in Nm	2	`OTG_Ctrl` in °
1	20	`OTG_Ctrl` in mm	10	`OTG_Ctrl` in mm	—	—
2	—	—	—	—	—	—
...	...	...	...	...	...	...
$m-1$	—	—	—	—	—	—

ID $l \in \{0, \ldots, (m-1)\}$. A set-point ${}^{l}_{k}D^{c}_{i}$ with $l = 0$ represents a desired robot state of first choice. $l = 1$ denotes the first alternative set-point and so on. In y-direction, force/torque control (`F/T_Ctrl`) with a desired value of $(-15\,N)$ has been chosen. If the module `F/T_Ctrl` is not available, for example, because there is no contact between the robot and its environment, the corresponding DOF will automatically be controlled by a distance controller (`Distance_Ctrl`) with a set-point of $0\,mm$. If this controller is also not able to service the DOF, for example, for the case that the robot is outside of the sensor's measurement range, the determined visual servo controller (`Vision_Ctrl`) is activated. If no further alternative is available, a save default backup controller, the open-loop pose controller, that is, the OTG algorithm (`OTG_Ctrl`), will be activated to guide the system safely to a defined state of motion. The OTG open-loop controller is the only module that is available at any instant, and that can act on arbitrary input values (cf. eqn. (3.17), p. 38). This important aspect on safety and stability will be further discussed in Sec. 7.4.

Of course, any combination of controllers and set-points is possible as a matter of principle. By means of the controller ID c, it is possible to choose the optimal controller for each task, that is, the most appropriate force/torque or visual servo concept, etc., is exerted. The necessary switching processes are handled by the *adaptive selection matrix*, which will be outlined in Sec. 7.2.2.

In contrast to common control interfaces, the MP not only allows the definition of desired values and a subset of parameters; furthermore, it enables the adaptation of the control system to the current system and environment state, such that it becomes possible to take care of the varying sensor and controller demands during robot task execution. Instead of using only one particular force/torque control scheme, it becomes possible to provide a set of force/torque controllers with different properties, such that — depending on the task — the most adequate controller can be chosen. The same considerations apply to other kinds of open- and/or closed-loop controllers. As a result, it is possible to have more than one controller ready for the same sensor signal, that is, the number of controllers m can be greater than the number of sensor systems n.

Stop Condition λ_i

During the execution of an MP, the task parameters $\mathcal{MP}_i$ remain constant. The stop condition λ_i defines a *system state* after whose attainment the currently executed MP becomes terminated. It is a Boolean expression that is defined as:

$$\lambda_i := \mathcal{S} \longrightarrow \{true, false\} \, . \tag{7.13}$$

$\mathcal{S}$ is the set of available sensors and their corresponding filter functions.

Example 7.2. *Equation (7.14) gives an example of a user stop condition that would let the execution of an MP end if the force of the Task Frame's z-direction averaged over 20 milliseconds gets below* $-15\,N$ *after a timeout of five seconds, or if the origin of the manipulator's hand frame w.r.t. the world frame gets above* $40\,mm$ *and the absolute velocity of the Task Frame's origin w.r.t. the world frame's x-axis gets higher than* $75\,mm/s$*.*

$$\begin{aligned} \lambda_i = & \left(\overline{{}^{task}_{\;\;z}F}^{20\,ms} < -15\,N\right) \vee \Big(\left({}^{world}_{\;\;\;x}p^{hand} > 40\,mm\right) \\ & \wedge \left(\left|{}^{world}_{\;\;\;x}v^{task}\right| > 75\,mm/s\right)\Big) \vee \left(t \geq 5000\,ms\right) \end{aligned} \tag{7.14}$$

If the stop condition becomes true at instant T_j with $j \in \mathbb{Z}$, we will have new task parameters $\mathcal{MP}_{j+1}$ in the following control cycle. This sensor-dependent change of parameters (cf. eqns. (7.2) – (7.10)) may also induce controller switchings in system depicted in Fig. 7.1.

Remark 7.4. *In robot programming paradigms widely used up to now, the cessation of motion is implicitly defined by the* move *command. In the MP programming paradigm, the end of a single motion command ($\mathcal{HM}_i$) is separated from the motion description, such that a higher flexibility for the specification and execution of sensor-guided and sensor-guarded robot motions can be achieved.*

7.2.2 Control Scheme for the Execution of Manipulation Primitives

Having given the formal definition of MPs above, let us now outline the respective control scheme, in which the on-line trajectory generation algorithm proposed in this monograph plays a key role. As the involved sensors and controllers may vary from DOF to DOF and from MP to MP, the control structure has to be adapted instantaneously according to any given situation.

The control architecture of Fig. 7.1 is based on a hybrid switched-system control approach [31, 163]. Five control submodules are assumed to be available: force/torque control (`F/T_Ctrl`), visual servo control (`Vision_Ctrl`), distance control (`Distance_Ctrl`), pose (`OTG_Ctrl`), and velocity control (`Velocity_Ctrl`). Of course the set $\mathcal{C}$ with its m elements can be expanded by further physical variables and controllers, respectively. At first glance the proposed structure seems similar to common hybrid switched-system control structures. But in contrast to common systems, the selection matrix is not static during one robot command: It depends on the current robot and environment state. This basic requirement proposed by Finkemeyer [85] is used as a basis for all following derivations.

In the classic approach for hybrid switched-system control in robotic systems, selection matrices $\mathbf{S}_i^c$ are used for each single controller c. These matrices can be generated from the *Compliance Frame* of each available controller type. If the controllers act in state space, six-dimensional control variable vectors ${}^{pos}\vec{o}_i^{\,c}$, ${}^{vel}\vec{o}_i^{\,c}$, and ${}^{acc}\vec{o}_i^{\,c}$ are computed by each controller c. Depending on the controller, these control variable vectors contain either *absolute* or *relative* values. The transfer function of a force/torque controller, for example, generates relative values, while a trajectory-following motion guided by the OTG algorithm can also be exerted with absolute values.

The desired new state of motion, which is subsequently transformed into joint space, can be computed by

$$ {}^{r}\vec{o}_i = \sum_{c=0}^{m-1} \mathbf{S}_i^c \; {}^{r}\vec{o}_i^{\,c} \qquad \forall \; r \in \{pos, vel, acc\} \;, \tag{7.15} $$

where the following condition must be fulfilled:

$$ \sum_{c=0}^{m-1} \mathbf{S}_i^c = \mathbf{I} \;. \tag{7.16} $$

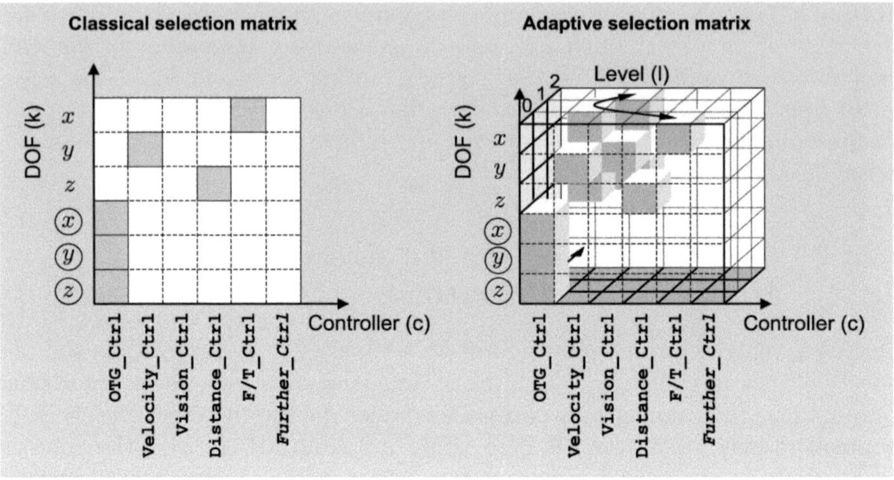

Fig. 7.6 Classical two-dimensional assignment (left) and adaptive three-dimensional assignment (right) of controllers and DOFs. The figure exactly accords to the one from p. 93 of [85].

I represents the identity matrix. This results in the required unique assignment of controllers to DOFs. The left part of Fig. 7.6 depicts such a sample assignment.

As mentioned in Sec. 7.2.1, the MP approach allows the definition of several alternative set-points (eqns. (7.7) – (7.10)), such that stable and efficient robot control is feasible for each individual DOF. For the execution of MPs, which allow the usage of any kind and any number of sensors, those classical two-dimensional selection matrices do not suffice. The definition of several alternative control loops requires extension by a third dimension representing the *control level* l with $l \in \{0, \ldots, (m-1)\}$, this three-dimensional compilation is illustrated in the right part of Fig. 7.6. The control loops of a DOF x can be shifted along the double-headed arrow (e.g., switching from force to visual servo control in Fig. 7.6). Thus, the active control variable can no longer be determined by constant selection matrices $\mathbf{S}_i^c$ of eqn. (7.15). The selection works dynamically and depends on two factors:

1. The currently available sensors and controllers as well as the current system state.
2. The assignment of controllers and control levels per DOF, that is, $\mathcal{HM}_i$.

In Fig. 7.1 the adaptive selection matrix is indicated by a dashed box. Formally, a controller c, for example, force/torque, pose, or velocity controller, receives a number of set-points ${}_k^l\Psi_i^c$, parameters ${}_k^l\Phi_i^c$ (eqn. (7.7)), the Task Frame parameters (eqn. (7.3)), and the assignment matrix $\mathbf{Z}_i^c$, which is generated from the respective DOF and level combinations (k, l) (cf. eqn. (7.7)). After the control algorithm of controller c has been executed, its output delivers two items: The availability matrix $\mathbf{F}_i^c$, and three controller output matrices containing a new state of motion for each controlled DOF: ${}^{pos}\mathbf{O}_i^c$, ${}^{vel}\mathbf{O}_i^c$, ${}^{acc}\mathbf{O}_i^c$. Depending on the controller, the controller output matrices can contain either absolute or relative values (cf. Fig. 7.5). These four matrices are the basis for the calculation of the control variable assignment matrices $\mathbf{G}_i^r$ with $r \in \{pos, vel, acc\}$ and the flag assignment matrix $\mathbf{E}_i$, which are used to determine the adaptive selection matrix $\mathbf{H}_i$ at instant T_i. All mentioned matrices and all calculation steps will be explained in the following.

Assignment Matrices $\mathbf{Z}_i^c$

As mentioned in the previous section, it is possible to determine alternatives for each set-point. Thus, each DOF k contains several control levels l. The assignment of the c-th controller to a control level l is formally represented by an assignment matrix $\mathbf{Z}_i^c$:

$$\mathbf{Z}_i^c = \begin{pmatrix} {}^0_x z_i & {}^0_y z_i & \cdots & {}^0_{ⓩ} z_i \\ {}^1_x z_i & {}^1_y z_i & \cdots & {}^1_{ⓩ} z_i \\ \vdots & \vdots & \vdots & \vdots \\ {}^l_x z_i & {}^l_y z_i & \cdots & {}^l_{ⓩ} z_i \\ \vdots & \vdots & \vdots & \vdots \\ {}^{(m-1)}_{\quad x} z_i & {}^{(m-1)}_{\quad y} z_i & \cdots & {}^{(m-1)}_{\quad ⓩ} z_i \end{pmatrix} \in \mathbb{B}^{m \times 6}, \tag{7.17}$$

$$\text{where } \mathbb{B} = \{0, 1\} .$$

Each column corresponds to a DOF and each row represents a control level, where m determines the maximum number of control levels. An entry of '1' assigns the DOF of the c-th controller to the control level l. The matrix corresponds to a vertical slice of the three-dimensional adaptive selection matrix representation of Fig. 7.6 (right). All m assignment matrices $\mathbf{Z}_i^c \; \forall \, c \in \mathcal{C}$ are input matrices for the control submodules and are implicitly defined by the set-point set $\mathcal{D}_i$.

Example 7.3. *The following assignment matrix $\mathbf{Z}_i^{\texttt{Velocity\_Ctrl}}$ has been defined for the open-loop velocity controller at time T_i:*

$$\mathbf{Z}_i^{\texttt{Velocity\_Ctrl}} = \begin{pmatrix} 0\,1\,0\,0\,0\,0 \\ 1\,0\,1\,0\,0\,0 \\ 0\,0\,0\,0\,0\,0 \end{pmatrix} . \tag{7.18}$$

Hence, the controller is selected to manage the y-direction at level 0 and to manage the x- and z-direction at level 1. All remaining DOFs are controlled by some other controller.

Controller Output Matrices ${}^{pos}\mathbf{O}_i^c$, ${}^{vel}\mathbf{O}_i^c$, and ${}^{acc}\mathbf{O}_i^c$

The control variables of one single controller c are summarized in up to three vectors ${}^{r}\vec{o}_i^{w}$ with $r \in \{pos, vel, acc\}$, where the *pos*-vector contains poses, the *vel*-vector velocities, and the *acc*-vector accelerations. It depends on the controller c whether all three output matrices are in use. For computing the adaptive selection matrix $\mathbf{H}_i$, the control variables are represented by the diagonal elements of the three controller output matrices ${}^{r}\mathbf{O}_i^c$ with $r \in \{pos, vel, acc\}$. The matrices for one single controller c are defined as

$$
{}^{r}\mathbf{O}_i^c = \begin{pmatrix} {}^{r}_{x}o_i^c & 0 & 0 & 0 & 0 & 0 \\ 0 & {}^{r}_{y}o_i^c & 0 & 0 & 0 & 0 \\ 0 & 0 & {}^{r}_{z}o_i^c & 0 & 0 & 0 \\ 0 & 0 & 0 & {}^{r}_{\textcircled{x}}o_i^c & 0 & 0 \\ 0 & 0 & 0 & 0 & {}^{r}_{\textcircled{y}}o_i^c & 0 \\ 0 & 0 & 0 & 0 & 0 & {}^{r}_{\textcircled{z}}o_i^c \end{pmatrix} \in \mathbb{R}^{6\times 6} \quad \forall\, r \in \{pos, vel, acc\} \;. \tag{7.19}
$$

Thus, the control variable vectors ${}^{r}\vec{o}_i^c$ with $r \in \{pos, vel, acc\}$ of the c-th controller can be derived from the matrices ${}^{r}\mathbf{O}_i^c$ with $r \in \{pos, vel, acc\}$ as follows:

$$
{}^{r}\vec{o}_i^c = diag\left({}^{r}\mathbf{O}_i^c\right) \qquad \forall\, r\; \{pos, vel, acc\} \;. \tag{7.20}
$$

Availability Matrix $\mathbf{F}_i^c$

To avoid the usage of control variables of inoperative control loops (e.g., force/torque control in free space), the validity of the output values is indicated by a flag. A value of one denotes a valid control variable and zero denotes an invalid control variable. The elements of the availability matrix $\mathbf{F}_i^c$ contain the mentioned flags of the c-th controller. For six DOFs, the availability matrix $\mathbf{F}_i^c$ is given by

$$
\mathbf{F}_i^c = \begin{pmatrix} {}_{x}f_i^c & 0 & 0 & 0 & 0 & 0 \\ 0 & {}_{y}f_i^c & 0 & 0 & 0 & 0 \\ 0 & 0 & {}_{z}f_i^c & 0 & 0 & 0 \\ 0 & 0 & 0 & {}_{\textcircled{x}}f_i^c & 0 & 0 \\ 0 & 0 & 0 & 0 & {}_{\textcircled{y}}f_i^c & 0 \\ 0 & 0 & 0 & 0 & 0 & {}_{\textcircled{z}}f_i^c \end{pmatrix} \in \mathbb{B}^{6\times 6} \;. \tag{7.21}
$$

The corresponding availability flag vector $\vec{f}_i^c$ of the c-th controller can be written as:

$$
\vec{f}_i^c = diag\left(\mathbf{F}_i^c\right) \;. \tag{7.22}
$$

Control Variable Assignment Matrices $\mathbf{G}_i^r$ with $r \in \{pos, vel, acc\}$

The calculation of the resulting command variables for the new state of motion, that is, ${}^{pos}\vec{o}_i$, ${}^{vel}\vec{o}_i$, and ${}^{acc}\vec{o}_i$, requires the mapping of the control values of all controllers to the corresponding DOF k *and* control level l. This is possible, as the definition of eqn. (7.7) ensures that only one controller can be chosen per DOF k and level l. The mapping of the control values is represented by the control variable assignment matrices $\mathbf{G}_i^r$ with $r \in \{pos, vel, acc\}$, and the calculation of $\mathbf{G}_i^{pos}$, $\mathbf{G}_i^{vel}$, and $\mathbf{G}_i^{acc}$ leads to eqn. (7.23).

$$\begin{aligned}\mathbf{G}_i^r &= \textstyle\sum_{c=0}^{m-1} \mathbf{Z}_i^c \; {}^r\mathbf{O}_i^c \\ &= \begin{pmatrix} {}^{0}_{x}g_i^r & {}^{0}_{y}g_i^r & \cdots & {}^{0}_{\textcircled{z}}g_i^r \\ {}^{1}_{x}g_i^r & {}^{1}_{y}g_i^r & \cdots & {}^{1}_{\textcircled{z}}g_i^r \\ \vdots & \vdots & \vdots & \vdots \\ {}^{l}_{x}g_i^r & {}^{l}_{y}g_i^r & \cdots & {}^{l}_{\textcircled{z}}g_i^r \\ \vdots & \vdots & \vdots & \vdots \\ {}^{(m-1)}_{\quad x}g_i^r & {}^{(m-1)}_{\quad y}g_i^r & \cdots & {}^{(m-1)}_{\quad \textcircled{z}}g_i^r \end{pmatrix} \in \mathbb{R}^{m\times 6} \quad \forall\, r \in \{pos, vel, acc\}\end{aligned} \tag{7.23}$$

Each column corresponds to a DOF k and each row to one particular control level l.

Adaptive Selection Matrix $\mathbf{H}_i$

To calculate the control value vectors ${}^r\vec{o}_i$ with $r \in \{pos, vel, acc\}$, which contain the input values in task space for the joint controllers (cf. Fig. 7.1), the correct row of $\mathbf{G}_i^r$ with $r \in \{pos, vel, acc\}$ must be chosen for each single DOF. This selection is realized by the adaptive selection matrix $\mathbf{H}_i$, which is defined as

$$\mathbf{H}_i = \begin{pmatrix} {}^{0}_{x}h_i & {}^{0}_{y}h_i & \cdots & {}^{0}_{\textcircled{z}}h_i \\ {}^{1}_{x}h_i & {}^{1}_{y}h_i & \cdots & {}^{1}_{\textcircled{z}}h_i \\ \vdots & \vdots & \vdots & \vdots \\ {}^{l}_{x}h_i & {}^{l}_{y}h_i & \cdots & {}^{l}_{\textcircled{z}}h_i \\ \vdots & \vdots & \vdots & \vdots \\ {}^{(m-1)}_{\quad x}h_i & {}^{(m-1)}_{\quad y}h_i & \cdots & {}^{(m-1)}_{\quad \textcircled{z}}h_i \end{pmatrix} \in \mathbb{B}^{m\times 6} \,. \tag{7.24}$$

The columns are assigned to the DOFs, and the rows denote control levels. The entries are either '1' or '0', where '1' selects the corresponding value. Of course, exactly one '1' per column exists. Thus, we obtain

$$\sum_{l=0}^{m-1} {}^{l}_{k}h_i \stackrel{!}{=} 1 \quad \forall\, k \in \left\{x, y, z, \textcircled{x}, \textcircled{y}, \textcircled{z}\right\} . \tag{7.25}$$

Transposing $\mathbf{H}_i$ and multiplying it with $\mathbf{G}_i^r$ with $r \in \{pos, vel, acc\}$ results in three symmetric matrices, whose diagonal elements contain the resulting control value vectors ${}^r\vec{o}_i$ with $r \in \{pos, vel, acc\}$. They can be written as

$$ {}^r\vec{o}_i = diag\left((\mathbf{H}_i)^T \ \mathbf{G}_i^r\right) \quad \forall\, r \in \{pos, vel, acc\} \ . \tag{7.26} $$

Please note that the elements of $\mathbf{H}_i$ have not been determined yet. The flag assignment matrix $\mathbf{E}_i$ is responsible for their calculation.

Flag Assignment Matrix $\mathbf{E}_i$

The matrix $\mathbf{H}_i$ depends on the m availability matrices $\mathbf{F}_i^c \ \forall\, c \in \mathcal{C}$ of the m controllers. By means of the m according assignment matrices $\mathbf{Z}_i^c \ \forall\, c \in \mathcal{C}$, they are mapped to the corresponding DOFs and levels. This results in the flag assignment matrix $\mathbf{E}_i$, which is computed by eqn. (7.27).

$$ \begin{aligned} \mathbf{E}_i &= \textstyle\sum_{c=0}^{m-1} \mathbf{Z}_i^c \ \mathbf{F}_i^c \\ &= \begin{pmatrix} {}_xe_i^0 & {}_ye_i^0 & \cdots & {}_{ⓩ}e_i^0 \\ {}_xe_i^1 & {}_ye_i^1 & \cdots & {}_{ⓩ}e_i^1 \\ \vdots & \vdots & \vdots & \vdots \\ {}_xe_i^c & {}_ye_i^c & \cdots & {}_{ⓩ}e_i^c \\ \vdots & \vdots & \vdots & \vdots \\ {}_xe_i^{(m-1)} & {}_ye_i^{(m-1)} & \cdots & {}_{ⓩ}e_i^{(m-1)} \end{pmatrix} \in \mathbb{B}^{m\times 6} \end{aligned} \tag{7.27} $$

Determination of $\mathbf{H}_i$ and ${}^r\vec{o}_i$ with $r \in \{pos, vel, acc\}$

The columns (DOFs) of $\mathbf{E}_i$ can be mapped to the corresponding columns of $\mathbf{H}_i$. The mapping law for each single DOF $k \in \{x, y, z, ⓧ, ⓨ, ⓩ\}$ is given in the following table. A '×' entry means *'don't care.'*

${}_ke_i^0$	${}_ke_i^1$	…	${}_ke_i^c$	…	${}_ke_i^{(m-1)}$	${}_k^0h_i$	${}_k^1h_i$	…	${}_k^lh_i$	…	${}_k^{(m-1)}h_i$
1	×	…	×	…	×	1	0	…	0	…	0
0	1	…	×	…	×	0	1	…	0	…	0
0	0	…	1	…	×	0	0	…	1	…	0
0	0	…	0	…	1	0	0	…	0	…	1

As can be seen from the table, only one controller is active per DOF, such that the table is in accordance with eqn. (7.25). The available controller with the lowest *levelID* will be activated. The table can be rewritten for all DOFs $k \in \{x, y, z, ⓧ, ⓨ, ⓩ\}$ as:

$$
\begin{aligned}
{}^{0}_{k}h_i &= {}_{k}e_i^0 \\
{}^{1}_{k}h_i &= \overline{{}_{k}e_i^0} \wedge {}_{k}e_i^1 \\
&\vdots \\
{}^{l}_{k}h_i &= \overline{{}_{k}e_i^0} \wedge \overline{{}_{k}e_i^1} \wedge \ldots \wedge {}_{k}e_i^c \\
&\vdots \\
{}^{(m-1)}_{k}h_i &= \overline{{}_{k}e_i^0} \wedge \overline{{}_{k}e_i^1} \wedge \ldots \wedge \overline{{}_{k}e_i^c} \wedge \ldots \wedge {}_{k}e_i^{(m-1)}
\end{aligned}
\tag{7.28}
$$

Thus, we know the adaptive selection matrix $\mathbf{H}_i$ and can calculate the output values of the control cycle at instant T_i by using eqn. (7.26)

$$
{}^{r}\vec{o}_i = diag\left((\mathbf{H}_i)^T \ \mathbf{G}_i^r\right) \quad \forall \ r \in \{pos, vel, acc\} \ .
$$

After the three vectors ${}^{r}\vec{o}_i$ with $r \in \{pos, vel, acc\}$ are calculated, we can compute the desired new state of motion for the control cycle at instant T_i: ${}^{BF}\mathbf{M}_i^{HF}$. Depending on the controller, all elements of ${}^{r}\vec{o}_i$ with $r \in \{pos, vel, acc\}$ can either be absolute or relative values. Absolute values can be directly transformed into joint space, while relative ones require the state of motion of the previous control cycle ${}^{BF}\mathbf{M}_{i-1}^{HF}$. As can be found in Fig. 7.1, the resulting command variables for the new state of motion in task space ${}^{BF}\mathbf{M}_i^{HF}$ are transformed into actuator space, such that they can act as input values for the lower-level joint controllers.

7.2.3 Remarks on the Availability Flag Vector $\vec{f}_i^c$

The availability flag vector $\vec{f}_i^c$ of a controller c decides, whether the control submodule is currently able to cope with the current system state or not. Thus, each element of $\vec{f}_i^c$ is determined by a function that maps state variables, sensor signals, or (sensor) events that are known and/or detected by the controller c to a Boolean value. One element of the c-th controller at instant T_i can be written as

$$
{}_{k}f_i^c := \mathcal{S} \longrightarrow \mathbb{B} \quad \text{with } k \in \{x, y, z, ⓧ, ⓨ, ⓩ\} \ , \tag{7.29}
$$

where $\mathcal{S}$ is the set of available sensor signals. This mapping function can be realized individually for each controller. It can be a simple constant (${}_{k}f_i^c = 1$), the result of a comparison with a threshold, or any other Gordian function. This feature can be used to handle critical situations. Some examples are:

- As long as the end-effector is not in contact with the environment, the force/torque control submodule(s) can set the flag to zero, and at the moment of contact detection, the module is immediately ready for operation.

- In case of internal controller errors, the flag can be set to zero, for example, due to a missing sensor signal or due to sensor signals that are outside of a sensor's metering range.
- If switching to a controller yields jumping jerk/acceleration/velocity command variable values, the flag can be set to zero.
- Visual servo controllers can use this flag if their respective vision system does not provide adequate information [96].
- It also may happen that a controller is inhibited, because of the Task Frame specification $\mathcal{TF}_i$; for example, a distance controller may only work correctly if one of the Task Frame axes is collinear to the measurement direction of the distance sensor.

The particular property of the OTG algorithm is, that all elements of its availability flag vector

$$\vec{f}_i^{\,\texttt{OTG\_Ctrl}} \equiv (1, 1, 1, 1, 1, 1) \qquad \forall\ i \in \mathbb{Z} \tag{7.30}$$

are equivalent to one at any time, because the algorithm is able to cope with any input values (cf. eqn. (3.17), p. 38). Even if no other controller is able to cope with the situation, the OTG algorithm can *always* generate continuous command variables for any state of motion in order to ensure a continuous and stable system behavior.

7.2.4 Remarks on Task Frame Switchings

It is an important advancement to enable Task Frame switchings at arbitrary (e.g., sensor-dependent) instants. During the execution of one single primitive, the Task Frame is anchored either to the world frame, to the base frame, to an external frame, or to the hand frame. At the transition from one MP $\mathcal{MP}_{i-1}$ to the next, $\mathcal{MP}_i$, that is, from one arbitrary control cycle T_{i-1} (the control cycle, in which the stop condition λ_{i-1} becomes true) to the next T_i, the parameters of the Task Frame $\mathcal{TF}_{i-1}$ may also switch arbitrarily to $\mathcal{TF}_i$. As a result, two different kinds of switching in the hybrid switched-control system of Fig. 7.1 are considered:

Switchings during one Manipulation Primitive
During the execution of one single MP, the input values for the hybrid switched-system, that is, the set-point set $\mathcal{D}_i$ and the Task Frame parameters $\mathcal{TF}_i$, remain constant. The elements of the assignment matrices $\mathbf{Z}_i^c\ \forall\ c \in \mathcal{C}$, which are generated from $\mathcal{D}_i$, are also constant during the execution of one single MP. Whether an element of the switching selection matrix $\mathbf{H}_i$ at instant T_i is set to one or to zero, solely depends on the availability flag vectors $\vec{f}_i^{\,c}\ \forall\ c \in \mathcal{C}$ (cf. Sec.7.2.3).

Switchings between two Manipulation Primitives

After the stop condition λ_{i-1} becomes true in the control cycle at instant T_{i-1}, new values for $\mathcal{D}_i$ and $\mathcal{TF}_i$ are available in the control cycle at instant T_i. If the parameters of the set-point set $\mathcal{D}_i$ change, new assignment matrices $\mathbf{Z}_i^c\ \forall\ c\ \in\ \mathcal{C}$ are calculated, such that the elements of the selection matrix $\mathbf{H}_i$ also change, but this behavior is comparable to that of the previous item.

Changes in the Task Frame parameters $\mathcal{TF}_i$, however, require a transformation of all m control submodules, that is, the states of all controllers must be transformed in order to sustain continuous control output values. For the case that filters for sensor signals are applied — before and/or subsequent to a controller — the filter states, also have to be transformed into the new coordinate frame in order to sustain continuous control output values.

Compared to filters and closed-loop controllers, which always possess a *state*, the OTG algorithm is state-/memoryless. This module does not need to be transformed; it only receives input values given w.r.t. the new Task Frame.

Example 7.4. *Let us give an intuitive every-day example of this kind of switching. Imagine we are driving a car, and we only use a compass for orientation. A switching of Task Frame parameters from $\mathcal{TF}_{i-1}$ to $\mathcal{TF}_i$ is comparable to an abrupt change of the compass card orientation. If we want to head for the same target, and if we are aware of this orientation change, we do not change our direction of driving. The same must happen to all control submodules in order to ensure continuous and feasible controller output values.*

If the closed-loop controllers, for example, the force/torque controller, receive an intended jump in their set-point signals at some instant T_i, these modules react deterministically and *immediately*, depending on their transfer function. The OTG algorithm is able to cope with any input values (as a matter of principle), such that we can freely switch the Task Frame parameters, even *during* a high-performance motion. This means, in particular, that we are able to perform a motion w.r.t. a moving Task Frame, we can switch to a fixed one at any time instant, and the resulting motion will be continuous. Thereby, it does not matter, whether the currently executed motion is a sensor-guided motion or a trajectory-following motion.

As already described in Chap. 2.4.3 (p. 26), the problem of orthogonality is very relevant. Compared to classic approaches [176, 227, 228], orthogonality is always guaranteed by the adaptive selection matrix in principle. The only problem that appears in this context, is the possible occurrence of *representation singularities* for the rotational degrees of freedom (cf. Sec. 7.2.1). But this problem of representation singularities can be bypassed as described in [139] or [202].

One final issue regarding the Task Frame shall be discussed here. In the proposed concept, we established the requirement that the transformation from the anchor frame ANC_i to the Task Frame $\vec{\theta}_i$ (eqn. (7.3)) remains constant during the execution of one single MP. Let us briefly discuss whether this is a necessary policy. Depending on the robot task, one may have the idea to introduce a sensor-dependent Task Frame pose, that is,

$$\vec{\theta}_i = {}^{ANC}\vec{P}_i^{TF} = \vec{f}(\mathcal{S}) \tag{7.31}$$

and thus

$$ {}^{ANC}\vec{V}_i^{TF} = \vec{f}(\mathcal{S}) \tag{7.32}$$
$$ {}^{ANC}\vec{A}_i^{TF} = \vec{f}(\mathcal{S}) \ , \tag{7.33}$$

such that we generally obtain

$$ {}^{ANC}\mathbf{M}_i^{TF} = \mathbf{f}(\mathcal{S}) \ . \tag{7.34}$$

The critical point here is, that a sensor signal (subset of $\mathcal{S}$) can occur twice in the control-loop. Thus, the stability of the overall system depends on the function f in eqns. (7.31) – (7.34). Concluding this issue: A sensor-dependent Task Frame pose $\vec{\theta}_i$ can lead to undesirable control behavior. A second idea could be to use a function of time

$$\vec{\theta}_i = {}^{ANC}\vec{P}_i^{TF} = \vec{f}(t) \tag{7.35}$$

for the variation of the Task Frame pose during one MP. In this case, the control behavior (and stability) depends on the function of time of eqn. (7.35). As a result, it can be advantageous for some robot tasks to apply a time-varying Task Frame pose $\vec{\theta}_i$, but in general the determination of the function f of eqn. (7.35) does not guarantee a stable overall system.

7.2.5 Remarks on the OTG Module

The Type IV OTG controller submodule in Fig. 7.1, `OTG_Ctrl` plays a special role. After all other control submodules $\mathcal{C} \setminus \{\texttt{OTG\_Ctrl}\}$ (cf. eqn. (7.10)) have calculated their controller availability flag vectors

$$\vec{f}_i^c \qquad \forall\, c \in \mathcal{C} \setminus \texttt{OTG\_Ctrl} \tag{7.36}$$

and output vectors

$$ {}^r\vec{o}_i^c \qquad \forall\, (c, r) \in \Big(\mathcal{C} \setminus \{\texttt{OTG\_Ctrl}\}\Big) \times \{pos, vel, acc\} \ , \tag{7.37}$$

we can determine the selection vector $\vec{S}_i$, that is, one of the input vectors of the OTG algorithm, by using the elements of the adaptive selection matrix

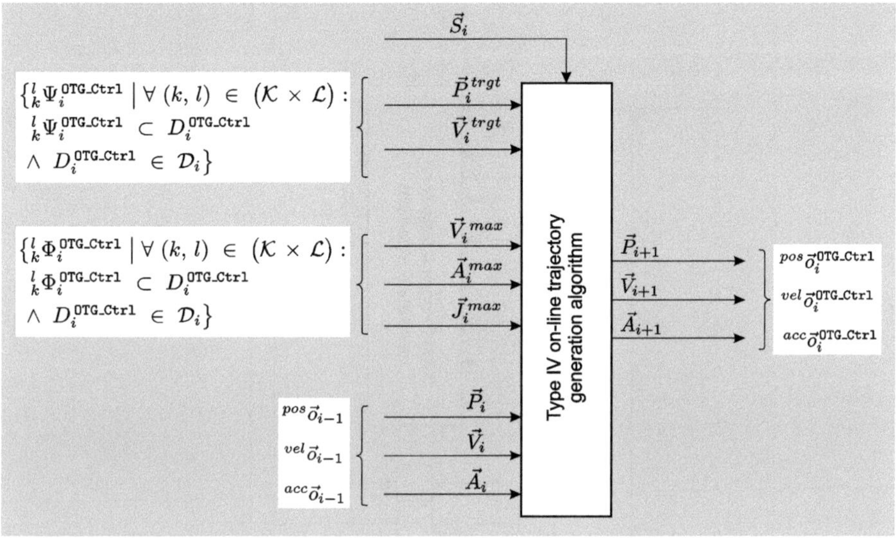

Fig. 7.7 Input and output values of the open-loop `OTG_Ctrl` module (cf. feedback loops in Fig. 7.1). This figure builds the bridge between Chap. 7 and Chaps. 3–6. Please also note the Remark 7.5.

$\mathbf{H}_i$ that refer to `OTG_Ctrl`. The motion constraints $\vec{V}_i^{max}$, $\vec{A}_i^{max}$, and $\vec{J}_i^{max}$ can be taken from the respective parameter values ${}^{l}_{k}\boldsymbol{\Phi}_i^{\mathtt{OTG\_Ctrl}}$ of $\mathcal{D}_i$, and the target state of motion, that is, $\vec{P}_i^{trgt}$ and $\vec{V}_i^{trgt}$, is determined by the set-points ${}^{l}_{k}\boldsymbol{\Psi}_i^{\mathtt{OTG\_Ctrl}}$ of $\mathcal{D}_i$, such that we obtain Fig. 7.7 to describe the relation of this chapter to Chaps. 3–6. When we recall eqn. (7.30)

$$\vec{f}_i^{\,\mathtt{OTG\_Ctrl}} \equiv (1,1,1,1,1,1) \qquad \forall\, i \in \mathbb{Z}\,,$$

it is of major importance, that eqn. (3.17) (p. 38) and all four criteria (i–iv) of Chap. 3.3 (p. 40) hold.

Remark 7.5. *How the time index i of the current state of motion* $\mathbf{M}_i$*, which is an input parameter of the OTG algorithm, is used, can be considered as a philosophical question:*

1. *We can either consider* $\mathbf{M}_i$ *as the current state of motion, and we calculate* $\mathbf{M}_{i+1}$ *for the next control cycle, or*
2. *we can consider* $\mathbf{M}_{i-1}$ *as the state of motion from the last control cycle, and the desired state of motion for the current control cycle* $\mathbf{M}_i$ *is calculated.*

Both points of views are correct. As in the field of control engineering the second notation is common, this chapter uses it, but in the rest of the book, the first notion is applied, because it greatly simplifies the comprehension of this work. Fig. 7.7 builds the bridge between these two parts, that is, between Chaps. 3–6 and this chapter.

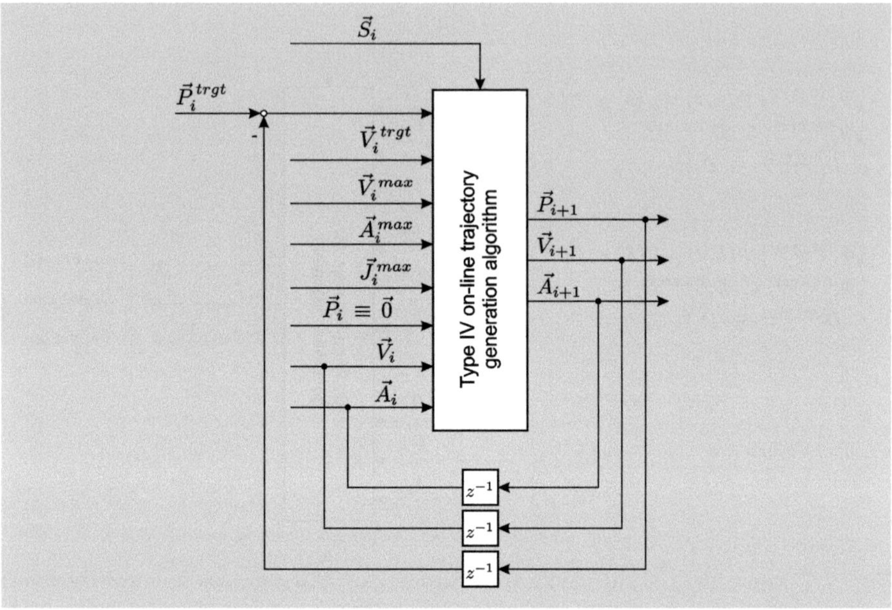

Fig. 7.8 Input and output values if the OTG algorithm is used for generating relative pose set-point values (cf. Fig. 7.5, p. 114).

As mentioned in Sec. 7.2.1 and shown in Fig. 7.5 (p. 114), computed motion set-points have to be interpreted either absolutely or relatively — depending on the currently executed Manipulation Primitive. If the output values of the OTG submodule are interpreted relatively, the input value $\vec{P}_i$ is equivalent to the zero vector, and the respective "feedback" value is subtracted from $\vec{P}_i^{trgt}$ as shown in Fig. 7.8.

The on-line trajectory generation submodule is responsible for the generation of all motions that are not sensor-guided. Furthermore, it is responsible for taking over control at any arbitrary time instant to assure a safe continuation of the robot motion — even if no other controller $c \in \mathcal{C}$ is ready for operation. Regarding Fig. 7.1, there is a further particularity for the OTG submodule: Its "feedback loop" is independent of the plant, but it behaves like a common closed-loop controller, although it is an open-loop one.

7.3 On-Line Trajectory Generation for Open-Loop Velocity Control

This section describes the open-loop velocity controller, that is, the controller submodule `Velocity_Ctrl` in Fig. 7.1. Although its development can be done in a straightforward way, it belongs to the completeness of this work to briefly introduce this control submodule.

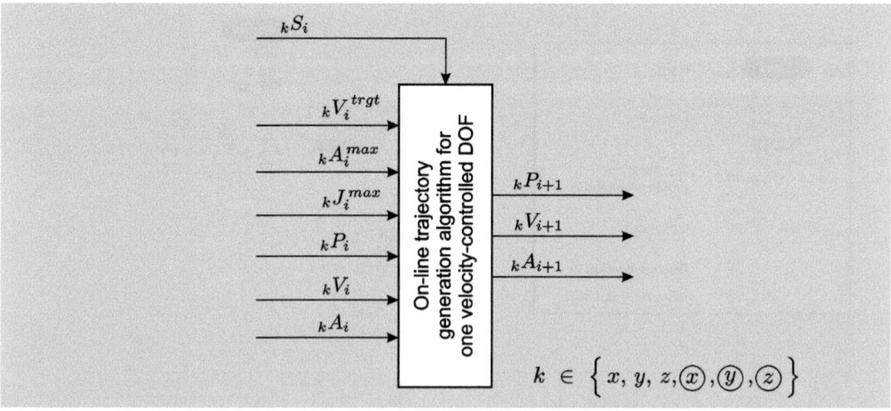

Fig. 7.9 Input and output values of the OTG algorithm for one single velocity-controlled DOF k (Types III – V).

The purpose of this module is to guide one single DOF k from any arbitrary state of motion to a certain velocity $_kV_i^{trgt}$. Regarding Table 3.1 (p. 39), the controller acts as a function

$$f:\ \mathbb{R}^{2\beta} \times \mathbb{B} \longrightarrow \mathbb{R}^{\beta}\ , \tag{7.38}$$

that is, three different open-loop velocity controllers can be developed: one with acceleration limitation (Types I – II in Table 3.1), one with jerk limitation (Types III – V in Table 3.1), and one with a limited derivative of jerk (Types VI – IX in Table 3.1).

This section only describes the open-loop velocity control module that corresponds to the OTG Types III – V. Fig. 7.9 shows its input and output values, and Fig. 7.10 depicts the complete decision tree, which is applied within this control module. We utilize six independent instances of the velocity control module of Fig. 7.9, which use the same motion state values $^{pos}\vec{o}_{i-1}$, $^{vel}\vec{o}_{i-1}$, and $^{acc}\vec{o}_{i-1}$ as the `OTG_Ctrl` one (cf. Figs. 7.1 and 7.7). If a DOF $k \in \left\{x, y, z, \textcircled{x}, \textcircled{y}, \textcircled{z}\right\}$ is guided by the open-loop velocity controller, its input values are comprised of the respective set-point $^l_k\Psi_i^{\texttt{Velocity\_Ctrl}}$ and parameters $^l_k\Phi_i^{\texttt{Velocity\_Ctrl}}$ of the set-point set $\mathcal{D}_i$ (cf. eqn. (7.7)):

$$\left\{^l_k\Psi_i^{\texttt{Velocity\_Ctrl}}\ \middle|\ \forall\,(k,l)\ \in\ (\mathcal{K}\times\mathcal{L}):\right. \tag{7.39}$$
$$\left.^l_k\Psi_i^{\texttt{Velocity\_Ctrl}} \subset D_i^{\texttt{Velocity\_Ctrl}} \wedge D_i^{\texttt{Velocity\_Ctrl}} \in \mathcal{D}_i\right\} \implies {_kV_i^{trgt}}$$

$$\left\{^l_k\Phi_i^{\texttt{Velocity\_Ctrl}}\ \middle|\ \forall\,(k,l)\ \in\ (\mathcal{K}\times\mathcal{L}):\right. \tag{7.40}$$
$$\left.^l_k\Phi_i^{\texttt{Velocity\_Ctrl}} \subset D_i^{\texttt{Velocity\_Ctrl}} \wedge D_i^{\texttt{Velocity\_Ctrl}} \in \mathcal{D}_i\right\} \implies \left({_kA_i^{max}},\, {_kJ_i^{max}}\right)\ .$$

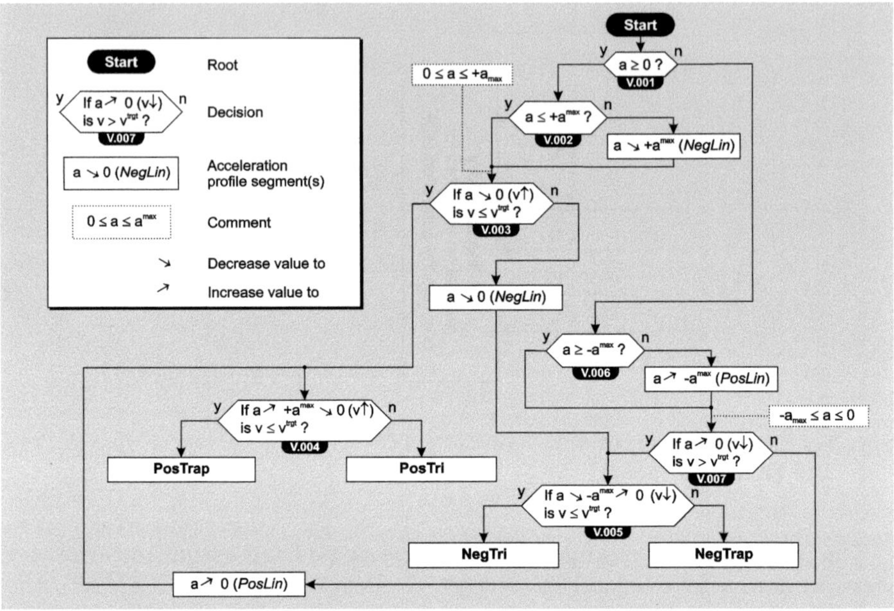

Fig. 7.10 Complete decision tree for velocity-controlled DOFs (Types III – V; required for the submodule `Velocity_Ctrl` in Fig. 7.1).

The resulting trajectory is composed in the same way as described in Chaps. 3 – 6. Up to $L = 5$ acceleration profile segments are connected to each other in order to time-optimally achieve the desired target velocity ${}_kV_i^{trgt}$. But in contrast to the OTG algorithms of the previous chapters, this control submodule does not generate time-synchronized, but only time-optimal trajectories for each selected DOF. To give an example of this control submodule, let us take the following arbitrary input values at $T_0 = 0\,ms$ for granted:

$$\begin{aligned}
{}_xP_0 &= 500\,mm & {}_yP_0 &= -200\,mm & {}_zP_0 &= 100\,mm \\
{}_xV_0 &= -400\,mm/s & {}_yV_0 &= 400\,mm/s & {}_zV_0 &= 50\,mm/s \\
{}_xA_0 &= 150\,mm/s^2 & {}_yA_0 &= -100\,mm/s^2 & {}_zA_0 &= -400\,mm/s^2 \\
{}_xV_0^{trgt} &= -200\,mm/s & {}_yV_0^{trgt} &= -200\,mm/s & {}_zV_0^{trgt} &= -200\,mm/s \\
{}_xA_0^{max} &= 250\,mm/s^2 & {}_yA_0^{max} &= 200\,mm/s^2 & {}_zA_0^{max} &= 500\,mm/s^2 \\
{}_xJ_0^{max} &= 300\,mm/s^3 & {}_yJ_0^{max} &= 500\,mm/s^3 & {}_zJ_0^{max} &= 200\,mm/s^3\ .
\end{aligned} \tag{7.41}$$

In particular the current states of motion

$${}_k\vec{M}_0 \;=\; ({}_kP_0,\, {}_kV_0,\, {}_kA_0) \qquad \forall\ k \in \{x, y, z\} \tag{7.42}$$

are arbitrary, and it is *essential*, that the algorithm covers the whole input space of $\mathbb{R}^6$ for any selected DOF. Fig. 7.11 presents the resulting trajectories for the input values of eqn. (7.41). The three translational DOFs reach their target velocities at the time instants:

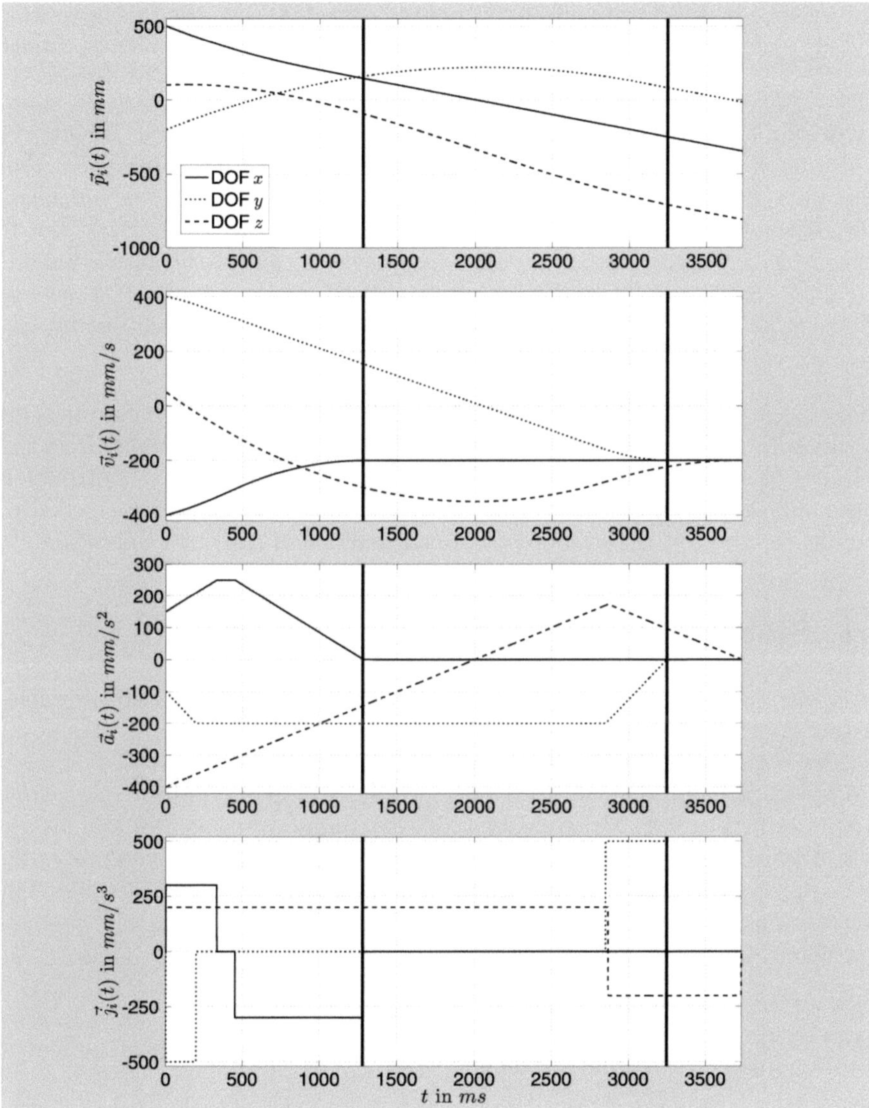

Fig. 7.11 Example of three trajectories generated by the open-loop velocity controller of Fig. 7.9. The input values of eqn. (7.41) were applied for their generation.

$$\begin{aligned} {}_x t_i^{min} &= 1.283\,s \\ {}_y t_i^{min} &= 3.249\,s \\ {}_z t_i^{min} &= 3.731\,s\ . \end{aligned} \tag{7.43}$$

This kind of command variable generation is useful for a huge set of robot tasks. If we want a manipulator to exert force/torque control in one DOF, and we know that the end-effector is not in contact, we can use this open-loop velocity controller to guide the robot into contact with a certain approach velocity.

Example 7.5. *A simple example is given by Table 7.1 (p. 115): The desired controller for the Task Frame's z-direction is the distance control submodule* `Distance_Ctrl`*, but for the case, that the distance controller is not available, because the sensor provides an out of range signal* ($_{z}f_{i}^{\texttt{Distance\_Ctrl}} = 0$*, cf. eqns. (7.21) and (7.22)), the open-loop velocity control submodule* `Velocity_Ctrl` *can guide the end-effector toward the environment with a velocity of* 15 *mm/s, and as soon as the distance sensor delivers a valid signal, the distance controller can take over control.*

Of course it would also be possible to add the feature of time-synchronization to this submodule, but — compared to the previously introduced OTG algorithms — it is not reasonable in practice, because the velocity controller is commonly applied as a transition controller, which becomes superseded as soon as the desired closed-loop control submodule is ready for operation.

7.4 Stability

After the previous formal introduction of the concept of hybrid switched-system control, which is in particular represented by Fig. 7.1, we are going to discuss the stability of such systems in this section.

This monograph does not provide a formal stability proof for the system presented in Fig. 7.1, which deals with the common problem of stability in hybrid switched-systems: Even if all control submodules behave stably, due to inappropriate switching sequences, the system can become unstable. The literature on such systems commonly distinguishes hybrid switched-systems into three subgroups [163]:

- State-dependent switching systems
- time-dependent switching systems, and
- autonomously switching systems.

The system to be regarded here is an autonomously switching system: The switchings occur in dependence on the set-point set $\mathcal{D}_i$ of the hybrid move command $\mathcal{HM}_i$ and on the stop condition λ_i of a single Manipulation Primitive $\mathcal{MP}_i$. Till now, there is no method that exists to prove the stability of this class of systems. For concrete set-ups, concrete environments, and concrete MPs, it is possible to apply multiple Lyapunov functions and Lie-algebraic stability criteria — but this is not a general solution [30, 31, 32, 163].

As a result, we can conclude that it is not yet possible to derive a proof of stability for the system depicted in Fig. 7.1. But of course we have to investigate this issue in order to legitimate the scheme of Fig. 7.1. The basic requirement for overall stability is a proof of stability for the control system in joint space, that is, the control cascade on the bottom right side of the dashed line in Fig. 7.1. This problem was surveyed very deeply, and we can find plenty of works on this topic [135, 196, 229, 246, 278]. Designing a robot joint control scheme is, of course, not a simple task, and there is *no golden rule* for it; commonly, a tradeoff between performance and robustness is required, but the treatment of this topic clearly goes beyond the scope of this book.

If we consider the joint space control scheme as stable, we can focus on the task space control scheme, which contains the hybrid switched-system. The thing we have to ensure is that if any control submodule is not able to control one or more DOFs of the system stably, we then have to provide a safe *backup controller*. This controller takes over control in the case of sensor malfunction or any other case that might lead one of the closed-loop controllers to become unstable. Here, the OTG algorithm comes into focus again, because this submodule is the only one that is able to cope with any state of motion, as we already stated in Chap. 5 and also through eqn. (7.30). Depending on the system, we can apply two different strategies:

The OTG algorithm takes over control in Cartesian space
This would mean, that we extend Table 7.1 by an additional control level m, such that we obtain $(m+1)$ control levels. At this control level, we *exclusively* apply the `OTG_Ctrl` module, either with a fixed set of input parameters, which would be the simplest solution, or with a set of on-line (state-dependently) calculated input parameters.

The OTG algorithm takes over control in joint space
The module *Open-loop position controller (Type IV OTG)* of the joint space control part of Fig. 7.1 guides all robot DOFs to a certain target position or state of motion. This target can again be either a set of pre-defined input parameters or a set of on-line computed ones.

In both cases, it is of major importance that the input parameters of the OTG algorithm are adequate at the moment of switching. Using either solution, the hybrid switched-system can be considered as a standard non-switched-system, such that the same methods as mentioned above [135, 160, 196, 229, 246, 278] can be utilized during the design and parameterizing procedure of the overall system. We can transform the problem of hybrid switched-system analysis to a stability analysis problem of a trajectory-following control scheme. The challenge and particularity of this analysis is that we have to consider an *arbitrary initial state of motion*: The state of motion that has been achieved by the control submodules, and that depends on sensor signals. Thus, the analysis result only depends on one parameter, that is, for a formal proof of stability, one would have to bound the allowed motion state, that is, the velocity and the acceleration of the end-effector. These joint space velocity and acceleration constraints are the same as the ones that have been used for the proof of stability of the joint space control scheme. This line of arguments leads us to the (expected and trivial) result that we have to ensure that the resulting velocities and accelerations that are computed in the *Transformation into joint space* block are limited.

To emphasize it again: This is not a formal proof of stability but only a logical line of arguments that leads to a heuristic and plausible result. Of course, this result is solely due to the OTG algorithm: *Before* the system becomes instable, this backup module takes over control either in Cartesian or in joint space and generates command variables for the system in any arbitrary state of motion.

7.5 Summary

A hybrid switched-system control scheme for robotic manipulation systems that enables the execution of sensor-guided *and* sensor-guarded robot motion commands was introduced. The OTG algorithm acts as one control submodule in this architecture and generates trajectories, that is, it is considered as an open-loop pose controller for one or more Cartesian DOFs. Furthermore, a second instance of the OTG algorithm is used in the same scheme and is able to take over control of all system DOFs in joint space in arbitrary states of motion.

The Manipulation Primitive framework, which was proposed by Finkemeyer [85], was formally introduced and extended to be suitable for state space control. Manipulation Primitives constitute an *interface* to hybrid switched-control systems in the field of robotic manipulation systems. Details of it were explained, and its relation to the OTG algorithms of the previous chapters was outlined. The most essential properties of this idea, which is basically represented by Fig. 7.1, are the following:

- Trajectory-following and sensor-guided robot motions can be arbitrarily combined, and we can sensor-dependently switch between these two modes, such that *sensor-guarded* robot motions (cf. Def. 1.1, p. 5) become realized.
- In particular, switchings from sensor-guided robot motion control to trajectory-following control become feasible.
- We can arbitrarily switch from joint space control to Cartesian space control or vice versa in arbitrary motion states and at unforeseen time instants.
- The pose and the anchoring of the Task Frame can be changed abruptly and sensor-dependently in arbitrary motion states and at unforeseen time instants.
- All mentioned switching procedures happen instantaneously within one control cycle, such that a kind of *robotic reflex* becomes enabled by this architecture.

The most important features of the OTG algorithm in this context are:

- It behaves as a closed-loop control submodule in a hybrid switched-system, but its "feedback" loop *bypasses* the plant, such that it is an open-loop controller that generates command variables for lower-level tracking control.

- It works independent of sensor signals (and sensor malfunctions).
- It can cope with arbitrary motion states.
- It can take over control even if no other control submodule is ready for operation.
- It can be utilized in joint space, such that all DOFs are guided by the algorithm, and it can work in Cartesian space, in which it guides a certain number of DOFs.

Example applications that underline the potential of all these properties can be found in the next chapter.

Chapter 8
Experimental Results and Applications

Chaps. 4 – 6 build the core contribution of this monograph, and Chap. 7 introduced an application case for the field of sensor-based robot motion control using on-line trajectory generation. While these four parts can be considered as formal and theoretical descriptions, we will discuss concrete experimental results in this chapter. The aim is to show how easily the OTG algorithm can be used in practice, and what new, essential possibilities of robot motion control suddenly become feasible.

8.1 Handling Arbitrary States of Motion

In order to start with a simple example, we assume a system with $K = 4$ DOFs being in an arbitrary state of motion. The Type IV OTG algorithm acts as function $f: \mathbb{R}^{32} \times \mathbb{B}^4 \longrightarrow \mathbb{R}^{12}$ (cf. Table 3.1, p. 39), which is computed every control cycle. The robot motion controller assumed in the following works at a frequency of $1\,KHz$, such that the algorithm is called once per $T^{cycle} = 1\,ms$. The given arbitrary motion parameters are (normalized values, without units):

$$\left.\begin{array}{l}\left.\begin{array}{rcl}\vec{P}_0 &=& (100, -200, 400, -800)^T \\ \vec{V}_0 &=& (300, -200, -50, 200)^T \\ \vec{A}_0 &=& (-350, -300, -50, 350)^T\end{array}\right\} \mathbf{M}_0 \\ \left.\begin{array}{rcl}\vec{P}_0^{trgt} &=& (-800, -500, -300, -400)^T \\ \vec{V}_0^{trgt} &=& (-50, -50, -100, -400)^T\end{array}\right\} \mathbf{M}_0^{trgt} \\ \left.\begin{array}{rcl}\vec{V}_0^{max} &=& (800, 750, 150, 600)^T \\ \vec{A}_0^{max} &=& (400, 400, 100, 300)^T \\ \vec{J}_0^{max} &=& (200, 400, 100, 600)^T\end{array}\right\} \mathbf{B}_0 \\ \begin{array}{rcl}\vec{S}_0 &=& (1, 1, 1, 1)^T\end{array}\end{array}\right\} \mathbf{W}_0\,. \qquad (8.1)$$

Fig. 8.1 discloses the resulting trajectory, and one can clearly recognize that all four DOFs reach their desired state of motion $\mathbf{M}_i^{trgt}$ at the same time

T. Kröger: On-Line Trajectory Generation in Robotic Systems, STAR 58, pp. 137–157.
springerlink.com

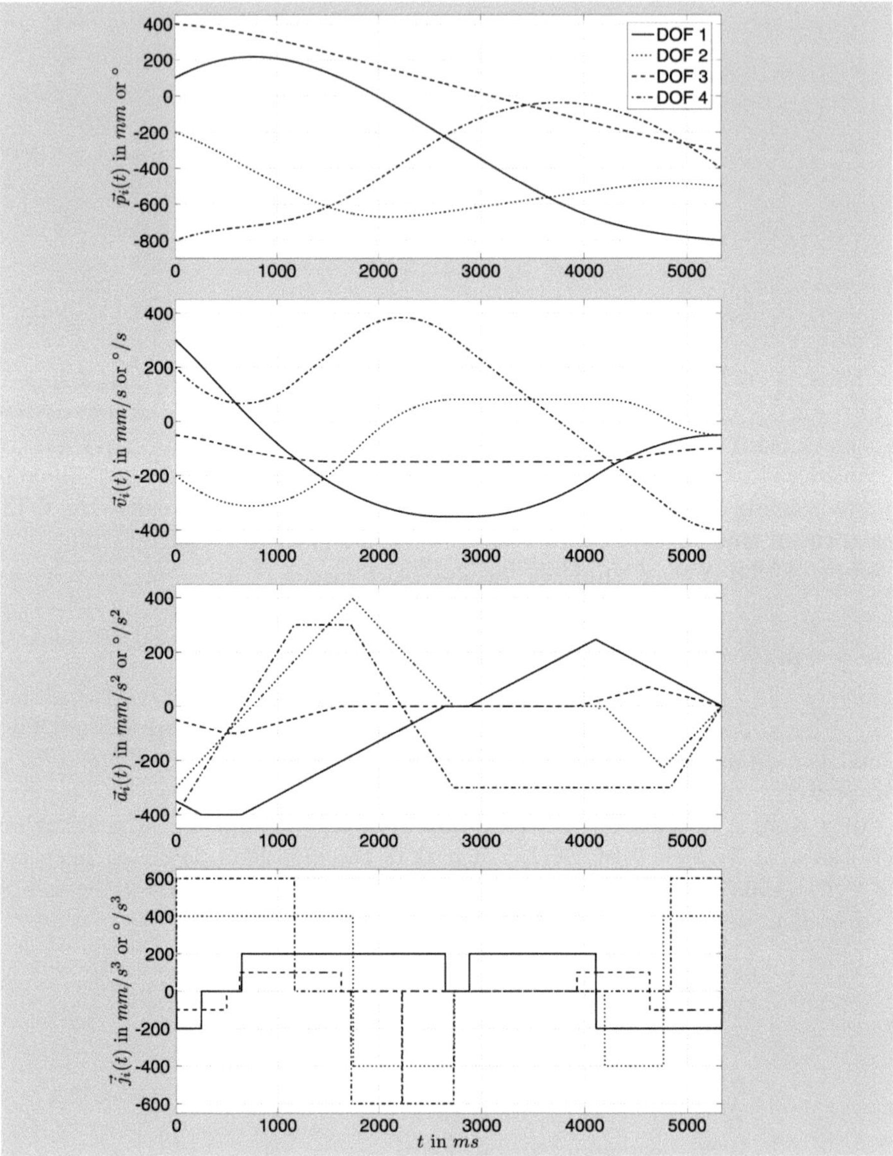

Fig. 8.1 Result of the Type IV trajectory for $K = 4$ DOFs, which was generated for the concrete input values $\mathbf{W}_i$ of eqn. (8.1).

instant $t_i^{sync} = 5340 ms \; \forall \; i \in \{0, \ldots, 5340\}$ ($N = 5340$ cycles, i.e the OTG algorithm was executed 5340 times). t_i^{sync} was determined by ${}_3t_i^{min}$ (DOF 3 reaches ${}_3V_i^{max}$, and the remaining three DOFs are synchronized on ${}_3t_i^{min}$). The selected acceleration profiles are:

$$\begin{array}{ll} {}_1\Psi_i^{Step2} = NegTrapZeroPosTri & {}_3\Psi_i^{Step2} = NegTrapZeroPosTri \\ {}_2\Psi_i^{Step2} = PosTriZeroNegTri & {}_4\Psi_i^{Step2} = PosTrapZeroNegTrap\ . \end{array} \quad (8.2)$$

For the sake of simplicity, the input parameters $\mathbf{M}_i^{trgt}$, $\mathbf{B}_i$, and $\vec{S}_i$ remained constant during the whole execution time from T_0 to T_N, that is, $\mathbf{M}_i^{trgt} = \mathbf{M}_0^{trgt} \wedge \mathbf{B}_i = \mathbf{B}_0 \wedge \vec{S}_i = \vec{S}_0 \ \forall\ i \in \{0, \ldots, 5340\}$.

8.2 Instantaneous Reaction to Unforeseen (Sensor) Events

The second example explains the basic methodology of how the OTG is applied to the simplest case of *sensor-guarded* motion control. For a simple and clear demonstration, we consider only a two-DOF Cartesian robot.

Fig. 8.2 presents the geometric *path* of a simple point-to-point motion (dotted line). It is the robot's task to move from an initial position $\vec{P}_0$ to a target position $\vec{P}_0^{trgt}$ under the constraint of boundary values $\vec{V}_0^{max}$, $\vec{A}_0^{max}$, and $\vec{J}_0^{max}$:

$$\begin{array}{rlrl} \vec{P}_0 &= (100, 200)^T\ mm & \vec{V}_0 &= (0, 0)^T\ mm/s \\ \vec{A}_0 &= (0, 0)^T\ mm/s^2 & \vec{V}_0^{max} &= (300, 200)^T\ mm/s \\ \vec{P}_0^{trgt} &= (800, 850)^T\ mm & \vec{A}_0^{max} &= (200, 300)^T\ mm/s^2 \\ \vec{V}_0^{trgt} &= (0, 0)^T\ mm/s & \vec{J}_0^{max} &= (400, 500)^T\ mm/s^3\ . \end{array} \quad (8.3)$$

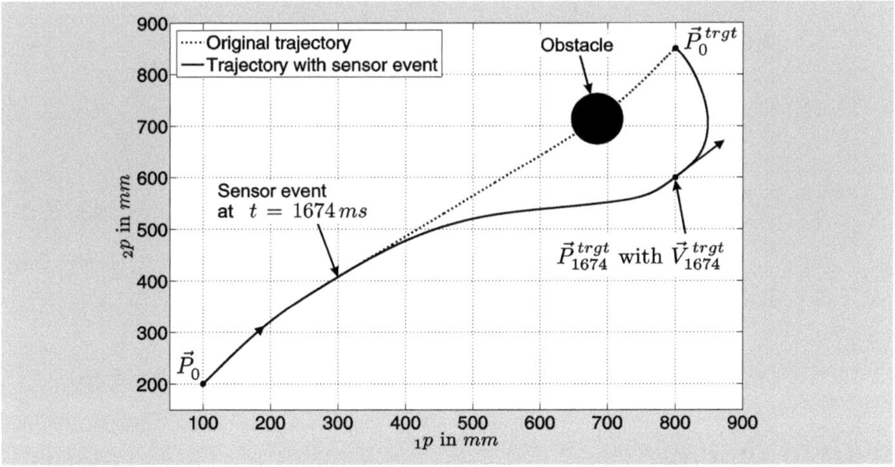

Fig. 8.2 XY-plot of the exemplary geometric *path* according to the trajectory shown in Fig. 8.3, with and without sensor event.

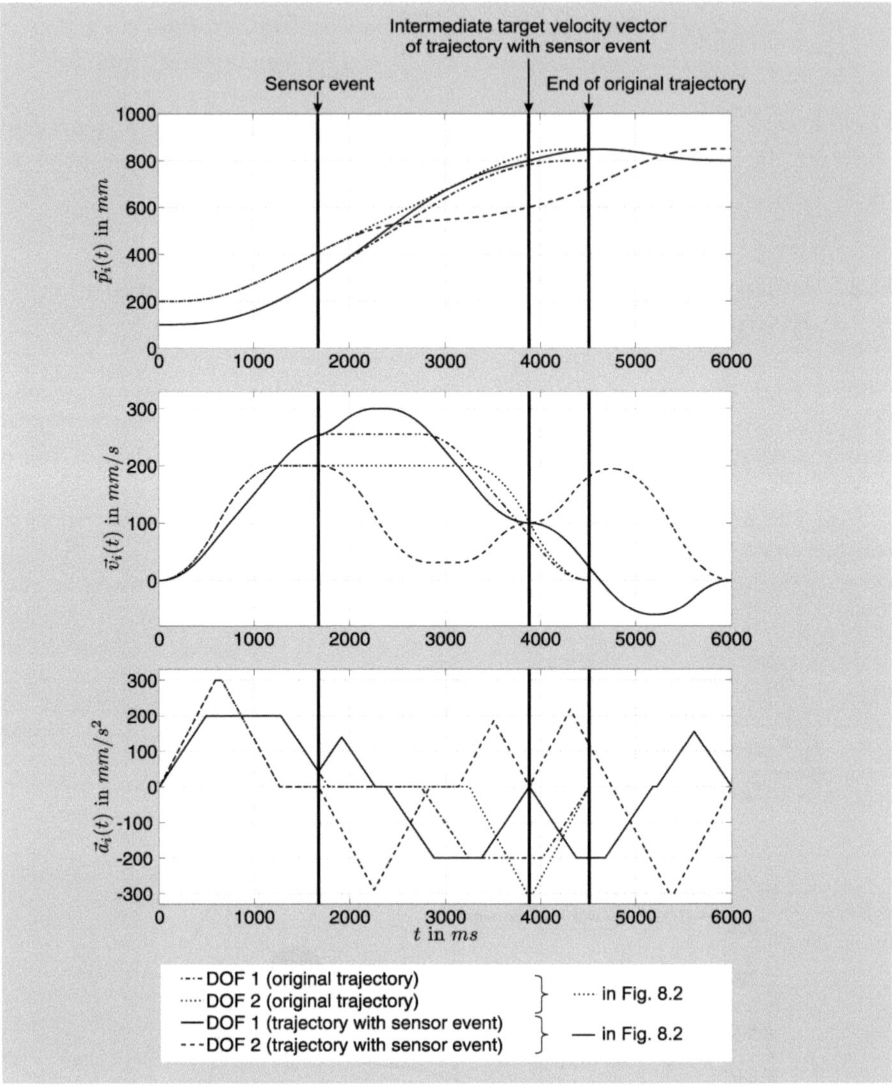

Fig. 8.3 Position, velocity, and acceleration progressions of the two-DOF Type IV *trajectory* that corresponds to the path of Fig. 8.2.

In the following, we analyze two scenarios: with and without obstacle.

Without Obstacle

If we did not have to react to sensor events, off-line methods as described in Chap. 2.4.2 (p. 17) could be applied. The position, velocity, and acceleration progressions of the original trajectory are depicted in Fig. 8.3 (dash-dotted and dotted lines). As can be seen in the bottom diagram of Fig. 8.3, both DOFs are transferred into their target positions by trapezoidal acceleration profiles, such that both reach their target state at $t = 4518\,ms$ with symmetrical velocity profiles.

With Obstacle and Reaction to Sensor Event

If we detect unknown objects/obstacles in our workspace, the OTG algorithm can react right after their detection. The results of this procedure are illustrated by the solid line in Fig. 8.2 and the solid and dashed lines of the diagrams in Fig. 8.3.

Due to some sensor (e.g., a distance sensor or a camera), the system detects the unforeseen obstacle at $t = 1674\,ms$ (${}_1P_{1674} = 300\,mm$, cf. Figs. 8.2 and 8.3). A simple solution would be an abrupt change of the target position values at the moment of obstacle detection, for example, to $\vec{P}_{1674}^{trgt} = (800, 600)^T\,mm$, such that we would prevent a collision with the object. After reaching $\vec{P}_{1674}^{trgt}$, the old target position $\vec{P}_0^{trgt}$ could be reused to reach the originally desired position.

Please pay attention to the velocity and acceleration progressions in Fig. 8.3. Compared to the idea above, an advanced and more dynamic version is presented: The system specifies an *intermediate target velocity vector* $\vec{V}_{1674}^{trgt} = (100, 100)^T\,mm/s$ into the position $\vec{P}_{1674}^{trgt} = (800, 600)^T\,mm$ in order to bypass the obstacle dynamically. Right after $\vec{P}_{1674}^{trgt}$ and $\vec{V}_{1674}^{trgt}$ have been time-optimally reached (at $t = 3879\,ms$, cf. Fig. 8.3), a second (abrupt) switching of input parameters occurs, and the system sets up the original parameters again ($\vec{P}_{3879}^{trgt} = \vec{P}_0^{trgt}$ and $\vec{V}_{3879}^{trgt} = \vec{0}$). These are finally reached at $t = 6$ seconds (cf. Fig. 8.3).

Remark 8.1. *The OTG algorithm is executed every millisecond, and the output values of the algorithm lead to continuous trajectories, which reach each desired state of motion* time-optimally *and* time-synchronized. *Furthermore, all boundary values, $\vec{V}_0^{max}$, $\vec{A}_0^{max}$, and $\vec{J}_0^{max}$, are kept during the whole trajectory. How the intermediate positions and velocity vectors such as $\vec{P}_{1674}^{trgt}$ and $\vec{V}_{1674}^{trgt}$ are calculated depends on the system above the OTG (cf. horizontal view in Chap. 2.5, p. 30)*

8.3 Homothetic Trajectories

This section presents results that correspond to Chap. 6 and compares a time-synchronized trajectory with a homothetic one. Figs. 8.4 and 8.5 display the corresponding position, velocity, acceleration, and jerk progressions for both trajectories, whose geometric paths are shown in Fig. 8.6. As in the previous sections, we start at $T_0 = 0\,ms$ and assume a cycle time of $T^{cycle} = 1\,ms$.

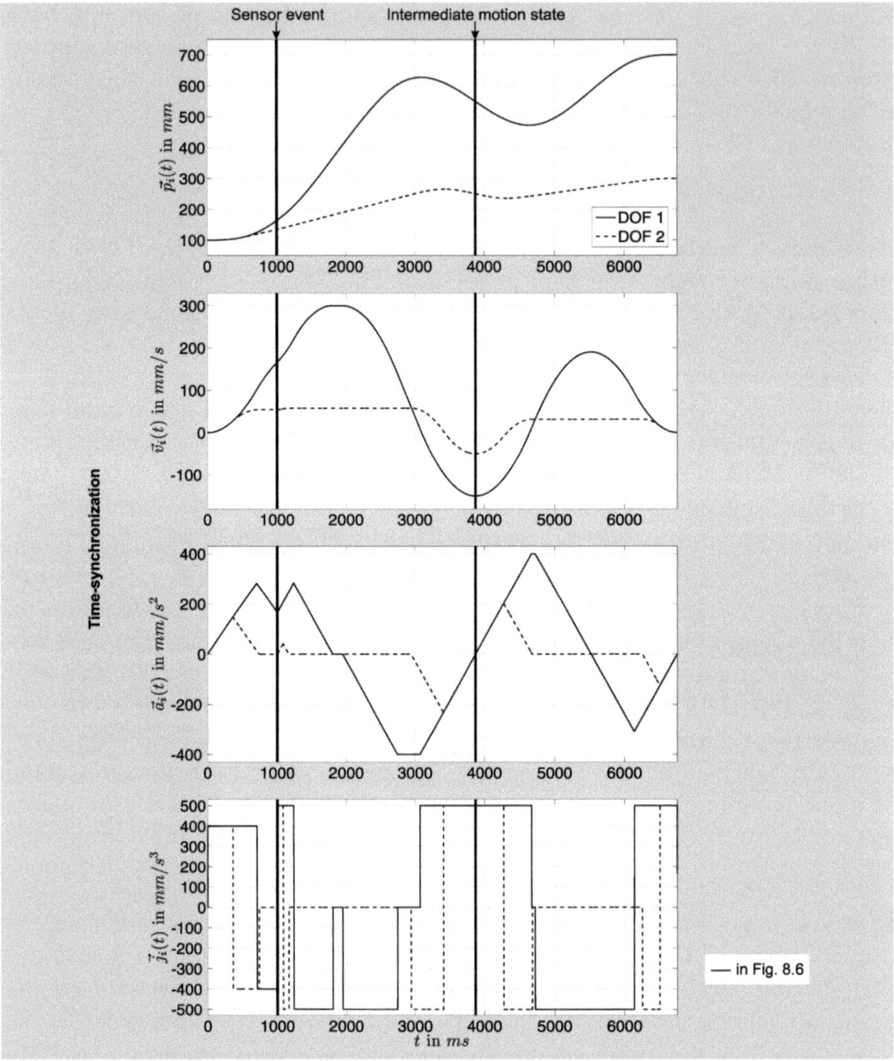

Fig. 8.4 Position, velocity, acceleration, and jerk progression of the *time-synchronized* two-DOF trajectory corresponding to the solid line in Fig. 8.6.

In this example, it is our task to execute the following motion from standstill:

$$
\begin{aligned}
\vec{P}_0 &= (100, 100)^T\,mm & \vec{V}_0^{max} &= (200, 200)^T\,mm/s \\
\vec{P}_0^{trgt} &= (700, 300)^T\,mm & \vec{A}_0^{max} &= (300, 300)^T\,mm/s^2 \\
\vec{V}_0^{trgt} &= (0, 0)^T\,mm/s & \vec{J}_0^{max} &= (400, 400)^T\,mm/s^3\ .
\end{aligned}
\tag{8.4}
$$

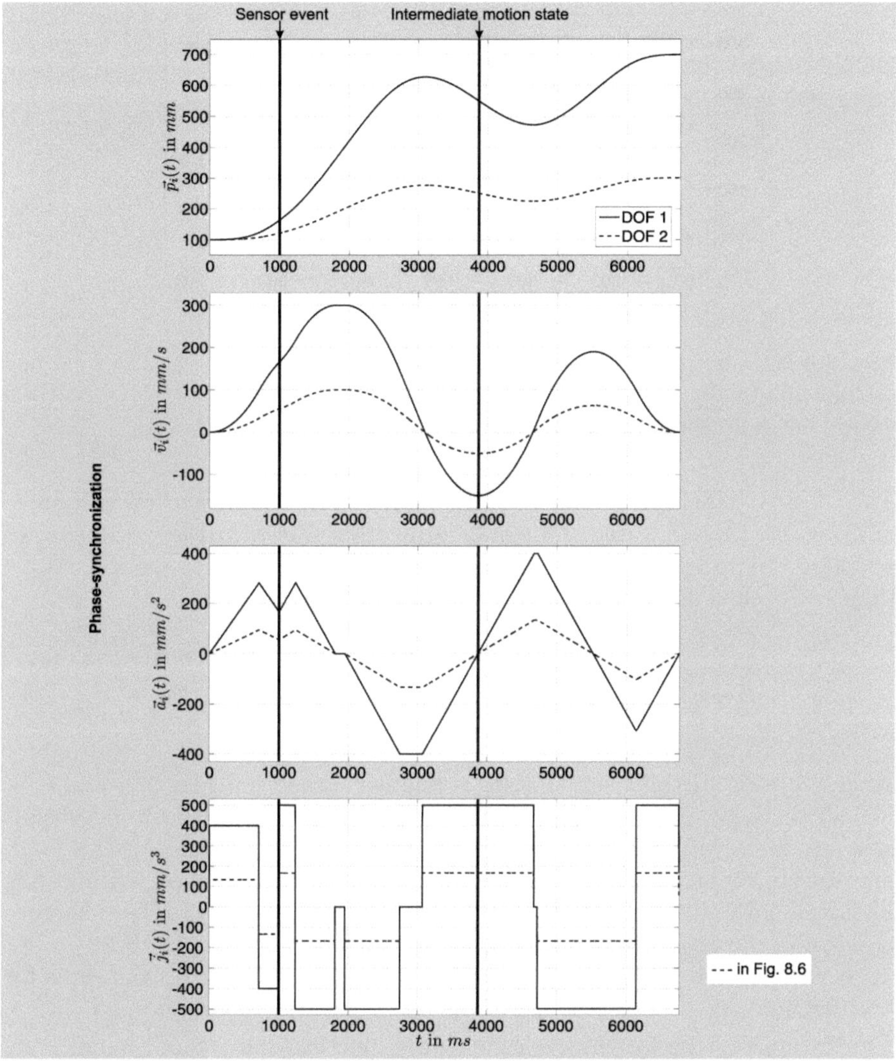

Fig. 8.5 Position, velocity, acceleration, and jerk progression of the *phase-synchronized* (homothetic) two-DOF trajectory corresponding to the dashed line in Fig. 8.6.

After $1000\,ms$, a sensor event happens (unforeseen), and due to this, we have to reach a position ahead with a negative velocity; in addition, the boundary values are arbitrarily changed:

$$\begin{aligned} \vec{P}_{1000}^{\,trgt} &= (550, 250)^T\,mm & \vec{V}_{1000}^{\,max} &= (300, 300)^T\,mm/s \\ \vec{V}_{1000}^{\,trgt} &= (-150, -50)^T\,mm/s & \vec{A}_{1000}^{\,max} &= (400, 400)^T\,mm/s^2 \\ & & \vec{J}_{1000}^{\,max} &= (500, 500)^T\,mm/s^3\ . \end{aligned} \tag{8.5}$$

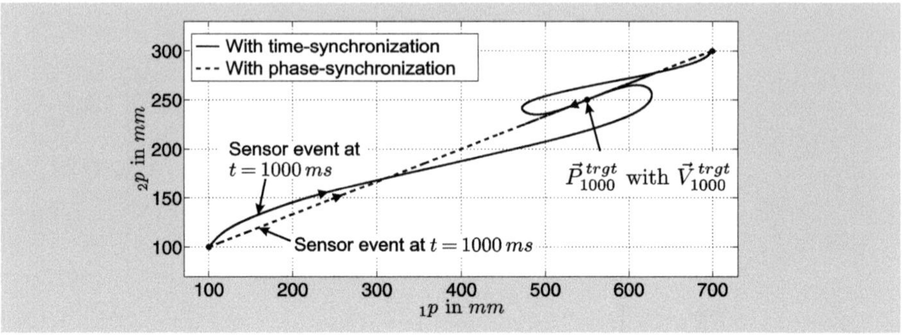

Fig. 8.6 XY-plot of the trajectories depicted in Figs. 8.4 and 8.5. The solid line corresponds to the *path* of the time-synchronized trajectory (Fig. 8.4), and the dashed line represents the *path* of the homothetic trajectory (Fig. 8.5).

This desired state of motion, $\mathbf{M}_{1000}^{trgt}$, is time-optimally reached after $t = 3873\,ms$ (cf. Figs. 8.4 and 8.5), and right after this a further switching of set-points happens, such that the originally desired state of motion, $\mathbf{M}_0^{trgt}$, shall be reached again:

$$\begin{aligned} \vec{P}_{3873}^{trgt} &= \vec{P}_0^{trgt} = (700, 300)^T\,mm \\ \vec{V}_{3873}^{trgt} &= \vec{V}_0^{trgt} = (0, 0)^T\,mm/s \end{aligned} \qquad \begin{aligned} \vec{V}_{3873}^{max} &= (300, 300)^T\,mm/s \\ \vec{A}_{3873}^{max} &= (400, 400)^T\,mm/s^2 \\ \vec{J}_{3873}^{max} &= (500, 500)^T\,mm/s^3\ . \end{aligned} \tag{8.6}$$

After $t = 6756\,ms$ this desired state is finally reached (cf. Figs. 8.4 and 8.5). For all three set-point sets, DOF 1 determines t_i^{sync}. Since both algorithms generate kinetically time-optimal trajectories, the progressions for this DOF are exactly the same in both cases, as can be verified in Figs. 8.4 and 8.5. In the homothetic case, DOF 1 is the reference DOF ($\kappa = 1$). The difference appears in the progression of all remaining DOFs, here only DOF 2.

As can be seen in Fig. 8.5, the acceleration profile is always the same for both DOFs: $_\kappa\Psi_0^{Step1} = PosTriZeroNegTri$ for the first 1000 cycles, $_\kappa\Psi_{1000}^{Step1} = PosTriZeroNegTrap$ for the cycles until $t = 3873\,ms$, and $_\kappa\Psi_{3873}^{Step1} = PosTrap\text{-}NegTri$ until $\mathbf{M}_{3873}^{trgt}$ is reached at $t = 6756\,ms$. Due to the case of homothety, it does not matter, whether we denote Step 1 profiles $_\kappa\Psi_i^{Step1}$ or Step 2 profiles $_\kappa\Psi_i^{Step2}$, since the profiles are the same for both steps. The determination of the profiles, however, is done in Step 1 (cf. Fig. 6.2, p. 102). The phase-synchrony can be observed in particular in the jerk diagrams of Figs. 8.4 and 8.5. While the time-synchronous case (Fig. 8.4) leads to asynchronous jerk progressions, which utilize zero or the values of $\vec{J}_i^{max}\ \forall\ i\ \in\ \{0, \ldots, 6756\}$ for both DOFs, the homothetic case (Fig. 8.5) applies the respectively adapted values $\vec{J}_i^{max\,'}\ \forall\ i\ \in\ \{0, \ldots, 6756\}$.

8.4 Unforeseen Switchings of Reference Coordinate Systems

As described in Chap. 7.2.1 (p. 111), robot control systems internally work with a number of different coordinate frames, whose motion states are updated every control period T^{cycle}. When a motion executed w.r.t. one of these frames becomes unexpectedly interrupted due to a sensor event, it may happen, that the motion command for the control cycle right after this sensor event is specified w.r.t. another frame, that is, the control system has to transform the whole state of motion (as well as the states of currently involved filters and closed-loop controllers) from one frame into another (cf. Chap. 7.2.4, p. 124). This leads to discontinuous internal motion state values, but the physical motion of course has to remain continuous. In fact, it has to be guaranteed, that the motion state progression is continuous!

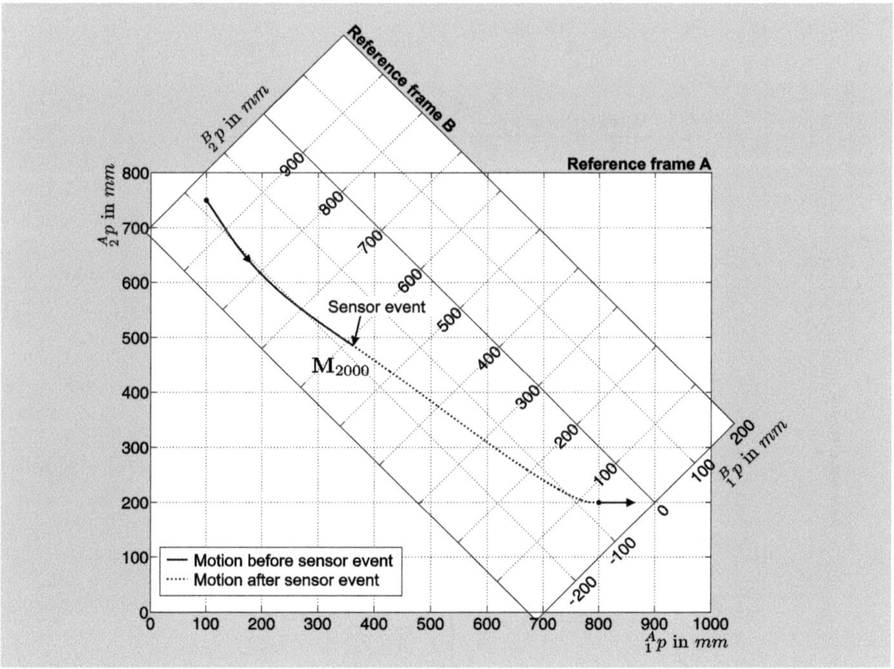

Fig. 8.7 XY-plot of a *path*, whose trajectory is executed w.r.t. reference frame A (solid line). Right after the indicated sensor event at $t = 2000ms$, frame B acts as the reference frame for the motion controller (dotted line).

Figs. 8.7, 8.8, and 8.9 depict a three-DOF trajectory, which is firstly specified w.r.t. reference frame A:

$$
\begin{aligned}
{}^{A}\vec{P}_{0} &= (100\,mm, 750\,mm, -30^{\circ})^{T} \\
{}^{A}\vec{V}_{0} &= (0\,mm/s, 0\,mm/s, 0^{\circ}/s)^{T} \\
{}^{A}\vec{A}_{0} &= (0\,mm/s^{2}, 0\,mm/s^{2}, 0^{\circ}/s^{2})^{T} \\
{}^{A}\vec{P}_{0}^{trgt} &= (800\,mm, 200\,mm, 90^{\circ})^{T} \\
{}^{A}\vec{V}_{0}^{trgt} &= (100\,mm/s, 0\,mm/s, 0^{\circ}/s)^{T} \\
{}^{A}\vec{V}_{0}^{max} &= (250\,mm/s, 400\,mm/s, 100^{\circ}/s)^{T} \\
{}^{A}\vec{A}_{0}^{max} &= (150\,mm/s^{2}, 250\,mm/s^{2}, 100^{\circ}/s^{2})^{T} \\
{}^{A}\vec{J}_{0}^{max} &= (600\,mm/s^{3}, 800\,mm/s^{3}, 400^{\circ}/s^{3})^{T}.
\end{aligned}
\tag{8.7}
$$

After a sensor event at $t = 2000\,ms$, the motion is not specified w.r.t. frame A anymore but w.r.t. reference frame B. This may have several reasons:

- The programmer or the higher-level system can only provide the new target state of motion w.r.t. frame B, for example, because the transformation ${}^{A}\mathbf{T}_{B}$ is time-variant and/or system-dependent.
- A frame, for example, the Task Frame (cf. Chap. 7.2.1, p. 109), abruptly changes its pose (due to the sensor event).

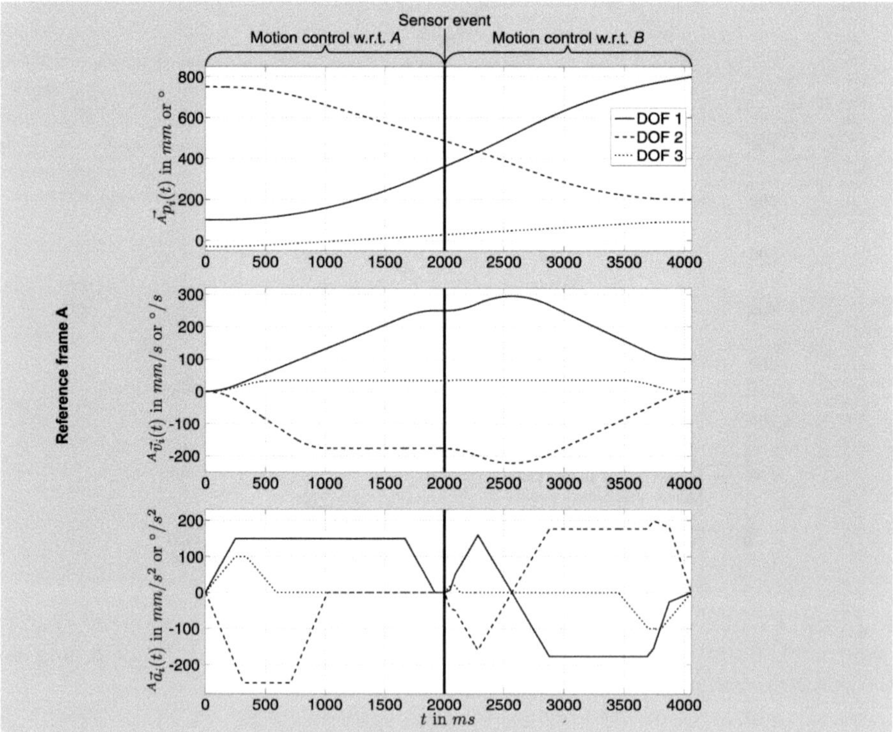

Fig. 8.8 Position, velocity, and acceleration progressions of the three-DOF trajectory of Fig. 8.7 w.r.t. reference frame A. The (unforeseen) sensor event happens at $t = 2000\,ms$.

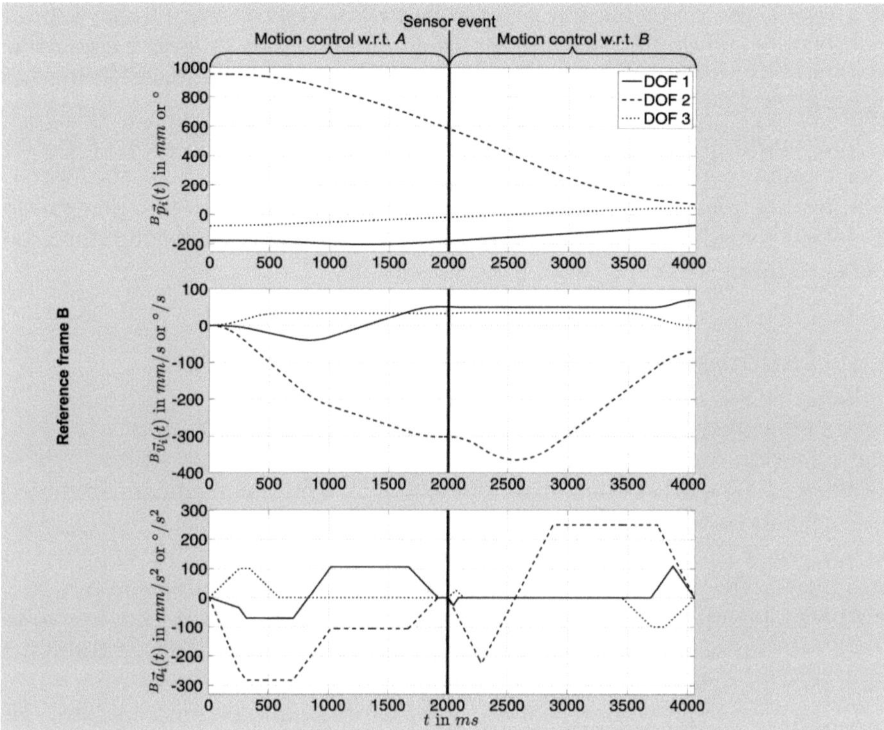

Fig. 8.9 The same trajectory as depicted in Fig. 8.8 but for the coordinates of reference frame B.

- Frame B was unknown at T_0.
- The change to frame B depends on the sensor event.
- The motion command before the sensor event is given w.r.t. a tool in system A, and right after the switching a tool located in B is used, such that the motion is specified with regard to this new tool.

The origin of frame B is located at position $(900, 200)\,mm$ w.r.t. frame A, and it is rotated by 45° (cf. Fig. 8.7). In this example, the target state of motion is transformed from A to B, and the absolute values of the kinematic constraints $\mathbf{B}_i$ remain constant (they could, of course, also be transformed):

$$\begin{aligned}
{}^{B}\vec{P}^{trgt}_{2000} &= (-70.7\,mm, 70.7\,mm, 45°)^T \\
{}^{B}\vec{V}^{trgt}_{2000} &= (70.7\,mm/s, -70.7\,mm/s, 0°/s)^T \\
{}^{B}\vec{V}^{max}_{2000} &= {}^{A}\vec{V}^{max}_{0} = (250\,mm/s, 400\,mm/s, 100°/s)^T \\
{}^{B}\vec{A}^{max}_{2000} &= {}^{A}\vec{A}^{max}_{0} = (150\,mm/s^2, 250\,mm/s^2, 100°/s^2)^T \\
{}^{B}\vec{J}^{max}_{2000} &= {}^{A}\vec{J}^{max}_{0} = (600\,mm/s^3, 800\,mm/s^3, 400°/s^3)^T.
\end{aligned} \tag{8.8}$$

As a result, the robot motion is controlled w.r.t. the reference frame B from $t = 2000\,ms$ on, that is, within the cycle of T_{2000}, the (a priori unknown) state of motion ${}^{A}\mathbf{M}_{2000}$ is transformed into the (a priori unknown) system of frame B.

The concept presented here can also be used for the specification of continuous motions consisting of several trajectory segments, such that the switchings are not sensor-dependent but pre-defined. Without on-line generation of motion control command variables, the here-presented cases of unforeseen reference frame switching would not be feasible.

8.5 Unforeseen Switchings of State Spaces

Very similar to what was presented in the previous section, it may happen, that a motion command that has to be executed right after a sensor event is specified in some other state space, for example, a motion command executed in Euclidian state space, is followed by one in joint state space (cf. Fig. 7.1, p. 106), or a Euler state space command is followed upon one in spherical coordinates. To realize this, we can apply the same procedure as in Sec. 8.4, but here all control values have to be nonlinearly transformed from one state space into another one within one single control cycle, that is, within the control cycle right after the sensor event.

To facilitate the comprehension, we take a simple r-φ-manipulator for demonstration. Fig. 8.10 shows the setup. The first task of this two-DOF r-φ-system is to move in Cartesian state space with the following motion parameters:

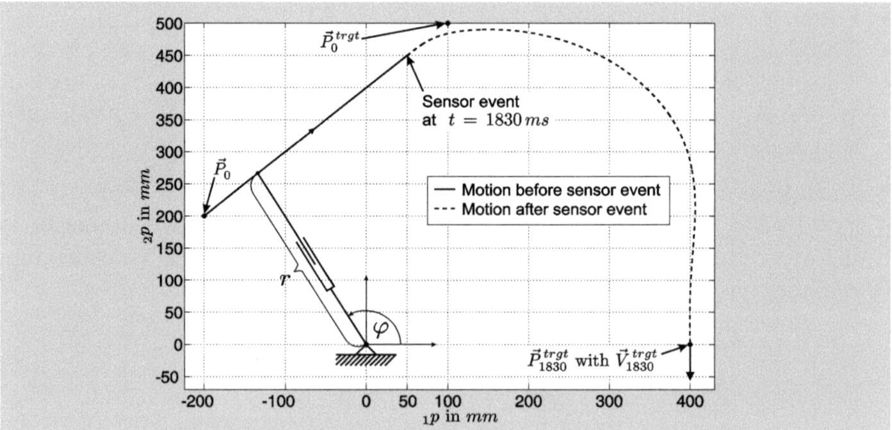

Fig. 8.10 r-φ-manipulator, which executes a Cartesian motion command that is unexpectedly interrupted at ${}^{Cart}\vec{p} = (50, 450)\,mm$ (solid line). Immediately after this event, the motion is controlled in joint state space until the target state of motion has been achieved (dashed line).

$$\begin{array}{llll} {}^{Cart}\vec{P}_0 = (-200,200)^T\, mm & {}^{Cart}\vec{V}_0^{max} = (300,300)^T\, mm/s & \\ {}^{Cart}\vec{P}_0^{trgt} = (100,500)^T\, mm & {}^{Cart}\vec{A}_0^{max} = (400,400)^T\, mm/s^2 & \\ {}^{Cart}\vec{V}_0^{trgt} = (0,0)^T\, mm/s & {}^{Cart}\vec{J}_0^{max} = (500,500)^T\, mm/s^3\ . \end{array} \tag{8.9}$$

We assume that a sensor event happens at $t = 1830ms$, the moment in which the manipulator has reached ${}^{Cart}\vec{p} = (50,450)\,mm$. The succeeding motion command is then specified in joint state space:

$$\begin{array}{ll} {}^{joint}\vec{P}_{1830}^{trgt} = (400\,mm, 0°)^T & {}^{joint}\vec{V}_{1830}^{max} = (150\,mm/s, 200°/s)^T \\ {}^{joint}\vec{V}_{1830}^{trgt} = (0\,mm/s, -20°/s)^T & {}^{joint}\vec{A}_{1830}^{max} = (200\,mm/s^2, 300°/s^2)^T \\ & {}^{joint}\vec{J}_{1830}^{max} = (800\,mm/s^3, 600°/s^3)^T\ . \end{array} \tag{8.10}$$

Fig. 8.11 shows the corresponding position, velocity, and acceleration progressions in Cartesian state space, and the same progressions in joint state space are presented in Fig. 8.12. The option to interrupt motion control in any state space, abruptly (and unexpectedly) switch to another state space, and steadily continue the motion in the new state space, provides us with the following new possibilities in robot motion control:

- If a motion cannot be continued in the current state space, for example, due to a priori unknown obstacles or singularities, switching to another state space can often bypass the problem very easily. This switching may happen at any time in any state of motion.
- If motion control parameters are abruptly changed, it can happen that the desired target state of motion can only be reached in another state space, for example, because of joint constraints.
- Switching from one state space to another at sensor-dependent time instants enables a very convenient and flexible way for robot motion specification with the possibility to automatically react to uncertainties.
- It is not always clear whether we have to switch the state space; but if we have to, which succeeding state space is the correct one? Since the motion set-points can be generated within one control cycle, the succeeding state space can now be sensor-dependently chosen immediately after the triggering event and does not have to be specified beforehand.

Although closely related to Sec. 8.4, the major difference to this example is that the transformation that is proceeded at a switching event must not necessarily be linear. As a matter of course, the resulting trajectories are again (kinetically) time-optimal and time-synchronized in the respective state space (cf. Figs. 8.11 and 8.12).

8.6 Hybrid Switched-System Control of a Six-DOF Industrial Manipulator

The results presented in the previous sections of this chapter were chosen to demonstrate different aspects of the basic functionality of the OTG algorithm.

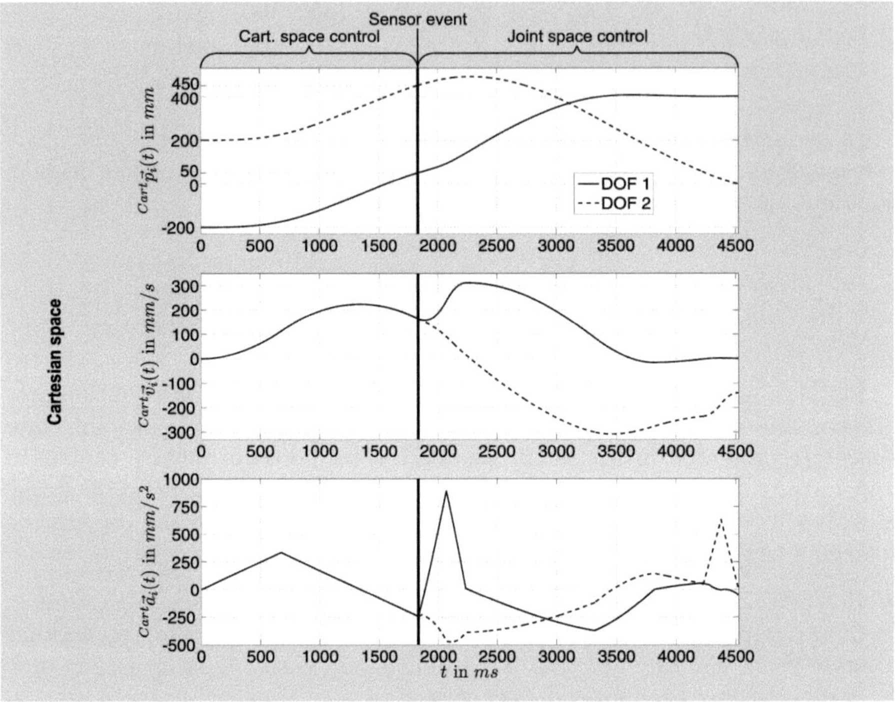

Fig. 8.11 Position, velocity, and acceleration progressions of the two-DOF trajectory of Fig. 8.10 in Cartesian state space. The sensor event, which leads to the unpredictable switching of the state space, happens in ${}^{Cart}\vec{p} = (50, 450)\,mm$ at $t = 1830\,ms$.

In order to demonstrate the functionality of the OTG concept in practice, this section presents results of real-world experiments.

Hardware Setup

Fig. 8.13 gives an overview of the utilized hardware setup. The original controller of a Stäubli RX60 industrial manipulator [250] was replaced, and the frequency inverters were interfaced. As shown in Fig. 8.13, a network of PCs with a QNX operating system [216] constitute the controller. The first PC interfaces the power electronics directly and is dedicated to position control in joint space (cf. dashed line in Fig. 7.1, 106). The second PC executes algorithms for hybrid switched-system control and acts as Manipulation Primitive interface to the third PC, which is dedicated to user applications. All PCs in this network communicate via the real-time middleware solution MiRPA (**Mi**ddleware for **R**obotics and **P**rocess Control **A**pplications, [86]), which is an in-house development of the Institut für Robotik und Prozessinformatik at the Technische Universität Braunschweig. MiRPA is a thin layer between the

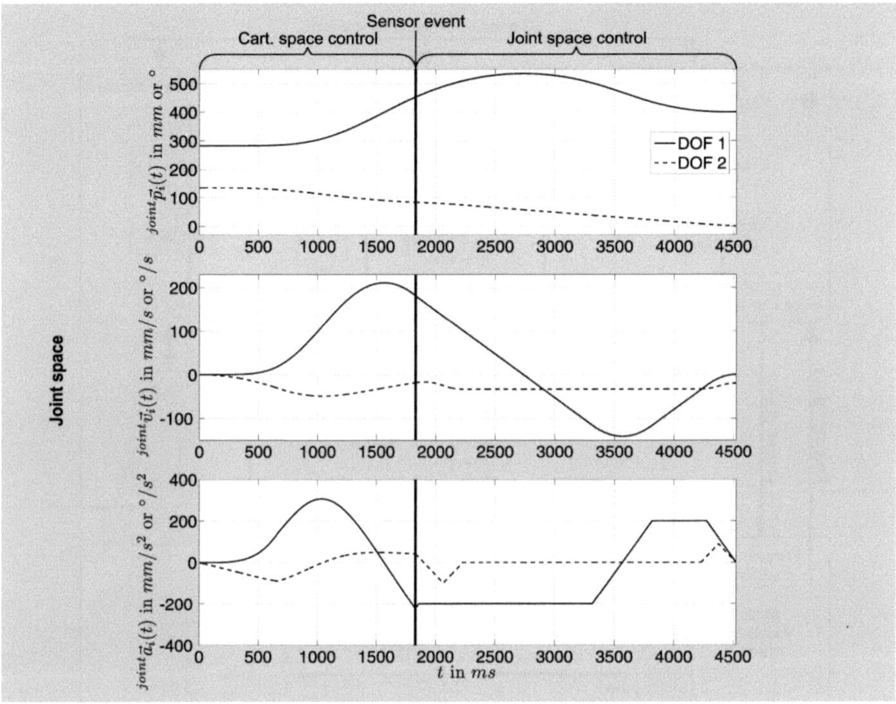

Fig. 8.12 Position, velocity, and acceleration progressions of the two-DOF trajectory of Fig. 8.10 in joint state space.

operating system and user processes and is responsible for distributed real-time interprocess communication. Its usage leads to the advantage that all processes only have one communication partner, and the system can be set up very modularly. Depending on the computational requirements, the number of MiRPA nodes can be chosen very freely, such that the number of nodes can be increased if the computational efforts grow. For the experiments shown in the following, a control rate of 10 KHz was applied for the joint controllers, and the hybrid switched-system controller runs at a frequency of 1 KHz.

Instantaneous Reactions to Unforeseen (Sensor) Events

To explain the behavior of instantaneous switchings from sensor-guided control to trajectory-following control, Fig. 8.14 now closes the loop of the introductory chapter (cf. Fig. 1.4, p. 4). At $T_0 = 0\,ms$, a sensor-guided robot motion command was executed w.r.t. the hand frame of the manipulator. All six Cartesian DOFs are controlled by a simple zero-force/torque controller (PID), which uses unfiltered force/torque values of a JR3 force/torque sensor

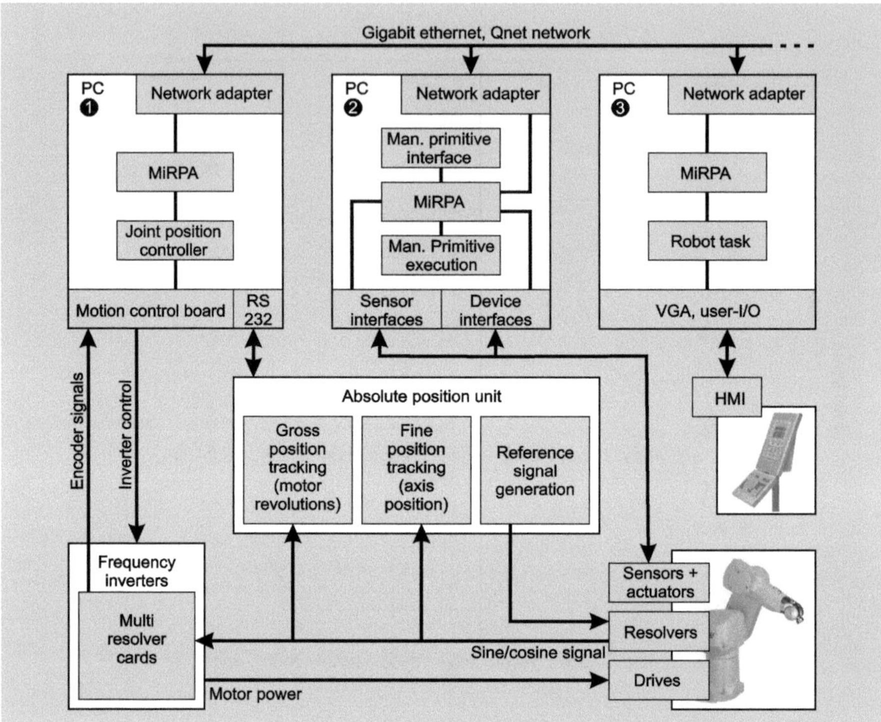

Fig. 8.13 Hardware architecture of the interfaced Stäubli RX60 industrial manipulator.

[119][1]. This was done intentionally in order to show the response of the overall system (including the OTG algorithm) on strongly noised sensor data. At $t = 586ms$, a (sensor) event happens, and the system instantaneously switches from sensor-guided robot motion control to trajectory-following control. The new trajectory is calculated instantaneously (within the control cycle after the event) and the manipulator performs a smooth, continuous motion. Here, this is done for all six DOFs $\{x, y, z, ⓧ, ⓨ, ⓩ\}$, that is, the selection vector switches from $\vec{S}_{585} = \vec{0}$ to $\vec{S}_{586} = \vec{1}$. Of course, it would also be possible that only some DOFs were switched from one controller to another. To describe the relevance for industrial practice: Imagine, a sensor (e.g., a force/torque sensor or a vision system) fails during a sensor-guided robot motion; the OTG algorithm can *always* take over control in any state of motion and at any time, such that a smooth, continuous motion results. Furthermore, if a desired (force/torque or vision) set-point cannot be achieved because of some reason, the current motion can be interrupted at any instant, such that the OTG algorithm guides respective DOFs to a safe state.

[1] Sensor model *85M35A-40 200N12*, receiver board running at 8 KHz.

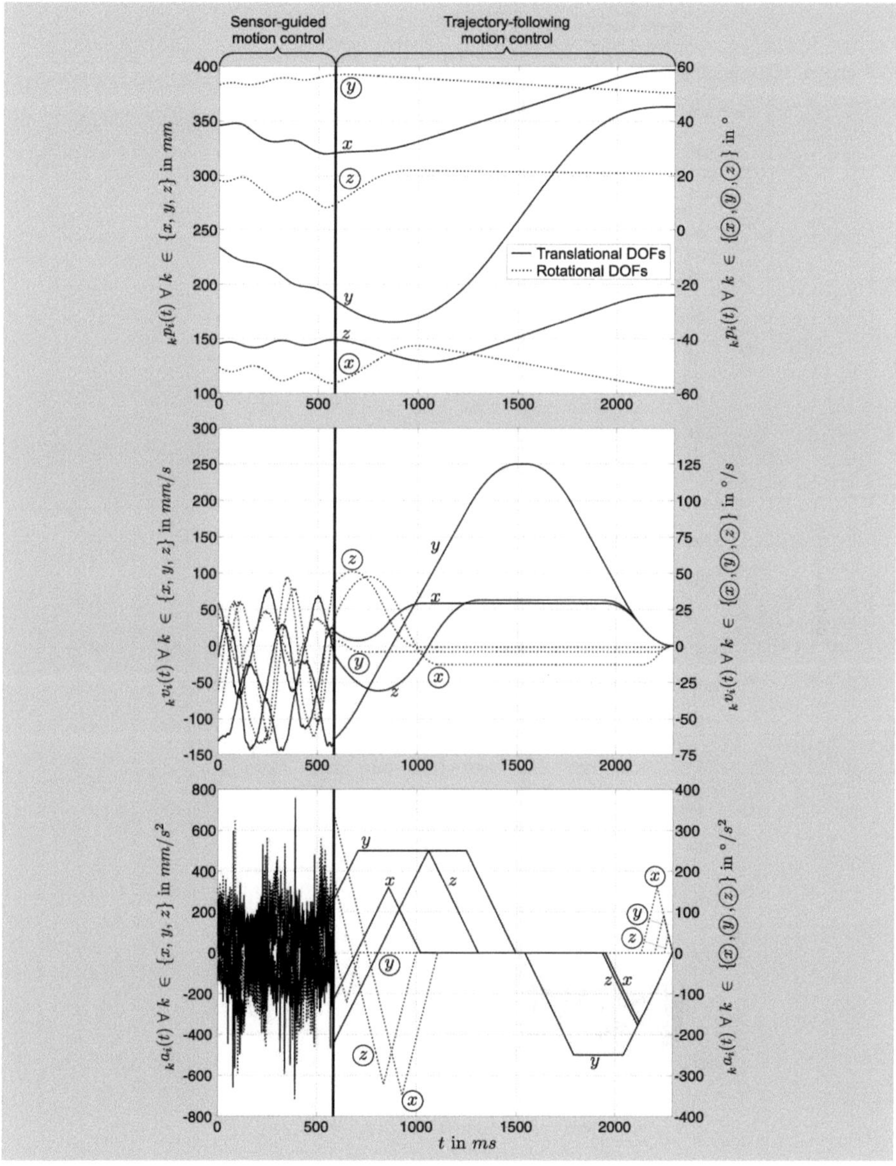

Fig. 8.14 Position, velocity, and acceleration progressions in Cartesian space of a sample motion of Stäubli RX60 industrial manipulator. First, all six DOFs are controlled by a feedback controller using force/torque sensor signals; at $t = 584\,ms$, an event happens, and the (open-loop) Type IV OTG algorithm takes over control.

Instantaneous Switchings from Task Space Control to Actuator Space Control

In addition to the previous subsection, we now extend the experiment and perform a switching from task space control to actuator space control.

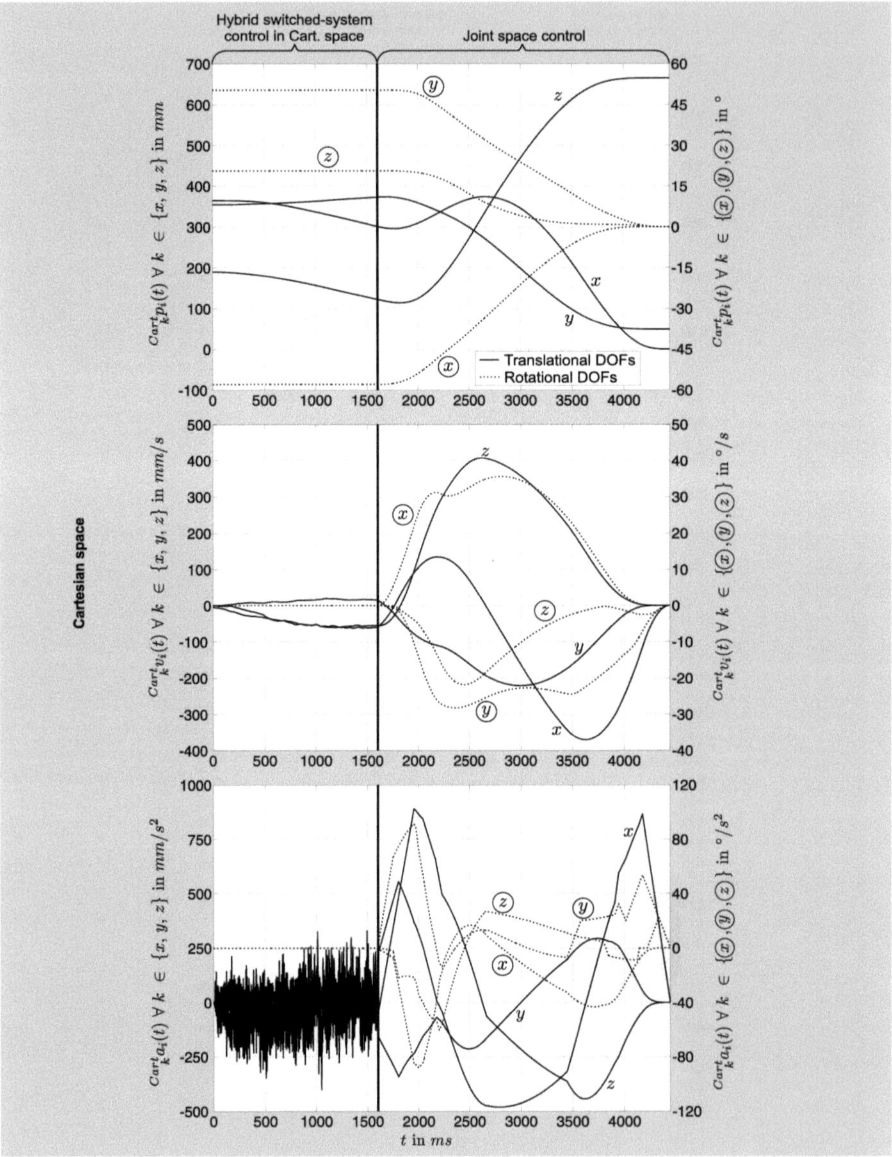

Fig. 8.15 Position, velocity, and acceleration progressions in Cartesian space of a sample motion of Stäubli RX60 industrial manipulator. (cf. Fig. 8.16).

Figs. 8.15 and 8.16 contain the results. The first figure shows a trajectory of all six DOFs in Cartesian space, and the second one illustrates the same trajectory in joint space. At $T_0 = 0ms$, a sensor-guided robot motion command was executed w.r.t. the hand frame of the manipulator ($ANC = HF$, cf. eqn. (7.5), p. 110). The three translational DOFs of the Task Frame are

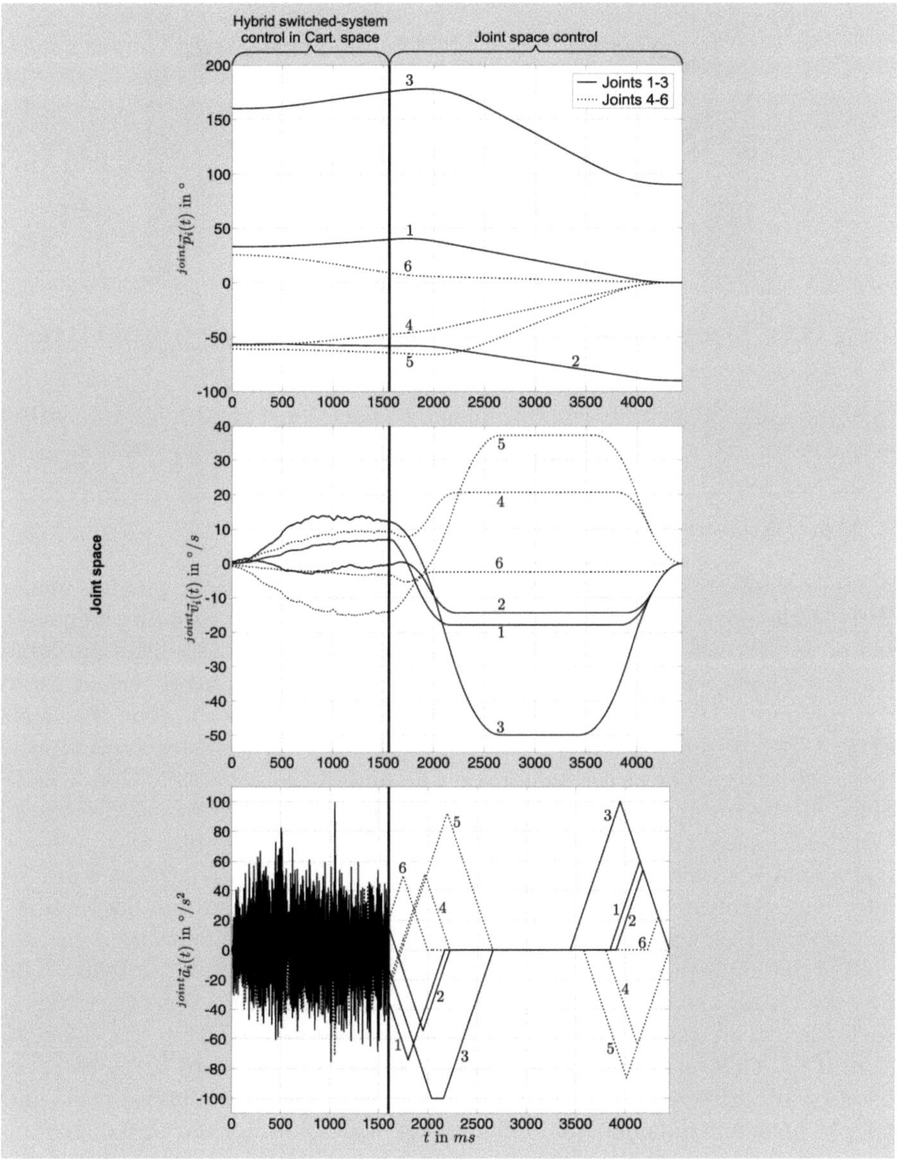

Fig. 8.16 The same trajectory as depicted in Fig. 8.15, but in joint space.

controlled by the same simple zero-force/torque controller as described in Sec. 8.6. The three rotational DOFs of the Task Frame are controlled by the OTG submodule, which only keeps their orientation. At $t = 1599\,ms$, a sensor event happens (the stop condition λ_0 becomes true, cf. Chap. 7.2.1, p. 116), and the system switches from task space control to joint space control, such

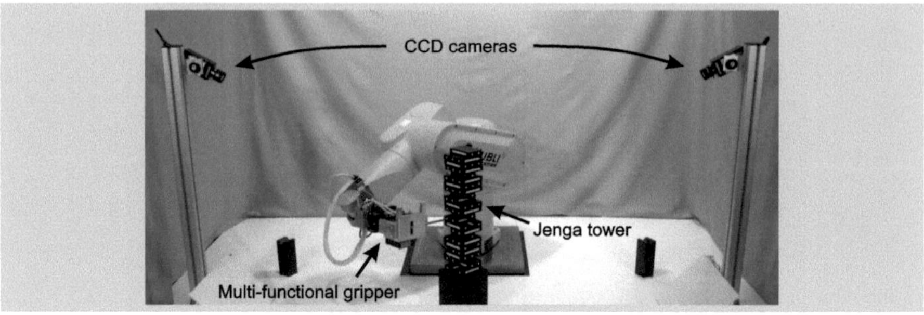

Fig. 8.17 Work cell set-up for the Jenga-playing manipulator [140, 141, 142].

that the OTG algorithm running in joint space (cf. Fig. 7.1, p. 106) takes over control and guides the manipulator to its initial (candle) position.

A More Advanced Application

A more challenging application is described in this section: playing Jenga [107]. Although it is not an industrial one, it is well suited to show the potential of multi-sensor integration and to demonstrate the possibilities available when concepts such as force/torque control, distance control, visual servo control, and OTG are combined. Actually, Jenga is a parlor game that consists of 54 wooden rectangular blocks. All blocks together are stacked as a tower with three blocks on each level. The aim of the game is to find a loose block in the tower, take it out, and put it back onto the top of the tower. With each move the tower becomes more and more unstable, and even for human players, this task requires a very high tactile sensitivity. The manipulator plays against itself with the aim of building a tower as high as possible; the set-up of its work cell is shown in Fig. 8.17. During the entire game, no strategy is applied; the blocks to be pushed out are selected randomly. The first step is always to try to push a block a few centimeters out of the tower, such that it subsequently can be gripped and pulled out from the opposite side. If the counter force during pushing becomes too high or if the cameras detect a dithering tower, the manipulator stops immediately, moves back, and tries to push out the next randomly chosen block. In order not to damage the tower when gripping a block, we have to ensure that the block to be gripped does not move when closing the gripper, that is, each block has to be gripped *exactly* centered. The spatial resolution of the three-dimensional model of the tower, which is estimated based on the CCD camera images, is approximately $1\,mm$ in the set-up. But this is not precise enough to calculate an accurate grip pose. This is the reason, why a triangulation distance sensor signal is additionally needed to measure the pose of a block in the range of micrometers. The distance sensor is mounted to the end effector of the manipulator. To perform a high-precision measurement of the pose of a single block, the manipulator moves its end-effector along the block, such that the distance sensor records a distance profile, which is subsequently used to determine the block's exact pose. When the block has been gripped, a force-guided MP is set up in order to pull the block out of the tower very carefully and to eliminate all transversal forces and respective torques. The last part of a single move is to put the block onto the top of the tower. This is done by a simple force-guarded MP, which moves the manipulator carefully towards the tower; after a certain force threshold is exceeded, contact has been established, the motion is stopped, and the gripper can be opened. Additional details and videos of this application can be found in [115, 142].

The majority of robot motion commands for the realization of this game are *sensor-guarded* motion commands (cf. Def. 1.1, p. 5), which were not executable this way without the OTG concept.

Chapter 9
Further Discussion

Before summarizing and concluding this book, some further marginal aspects are discussed in this chapter.

9.1 On-Line Trajectory Generation as an Interface to Non-Real-Time Systems

Robot control systems are often realized as PC- or microcontroller-based real-time systems, such that all low-level control algorithms are periodically executed in isochronous time intervals. Besides the advantages of the OTG approach described in Chap. 8, an additional benefit of the OTG algorithm, which was designed to be executed under real-time conditions, is that its input signals can act as an interface to non-real-time systems and/or systems with non-constant sampling intervals. Since we can react to arbitrary motion control set-point changes, the system can react to new values at any time instant in any state of motion $\mathbf{m}_i$. As a consequence, it becomes possible to process set-point values $\mathbf{M}_i^{trgt}$, $\mathbf{B}_i$, and $\vec{S}_i$ in time-varying intervals.

Computer vision systems, for example, often work with image processing periods T^{vision} that are relatively high compared to lower-level control rates for pose or force/torque control. Assuming that the sampling period of the hybrid switched-system of Fig. 7.1 (p. 106) is T^{cycle} again, it is possible that we cannot receive motion control set-points of the visual servo control submodule every cycle, but rather every ν-th cycle:

$$T^{vision} = \nu \; T^{cycle} \quad \text{with } \nu \in \mathbb{N} \backslash \{0\} \; . \tag{9.1}$$

In such a case, the OTG module can synchronously run on the low control level and provide set-points for the underlying joint controllers every control period T^{cycle}. The OTG input parameters for DOFs that are selected for visual servo control are only changed once per T^{vision}. Thereby, ν must not necessarily be constant. If the computer vision algorithms are realized on non-real-time platforms such as Linux or Windows, we usually have to consider a

T. Kröger: On-Line Trajectory Generation in Robotic Systems, STAR 58, pp. 159–177.
springerlink.com

more or less significant jitter, such that T^{vision} is non-constant. In this case, the OTG module can be used to provide a very convenient motion control interface to non-real-time systems.

9.2 Visual Servo Control

An overview of basic works on visual servo control was already presented in Chap. 2.4.3 (p. 26), and the previous section indicated that the OTG algorithm is suitable to be used as an intermediate layer between vision algorithms and robot motion control.

Visual servo control methods are commonly distinguished in position-based and image-based ones. In position-based concepts, image features are detected and used to estimate the camera position, whereas image-based concepts calculate an error signal that is derived from the image and map this signal directly to actuator command variables. The approach of Gans and Hutchinson [95, 96, 97] combines both methods in one hybrid switched-system and proves Lyapunov stability. The general aim in both solutions is to minimize an error function, which often leads to very unsmooth robot motion behavior, for example, due to lost features or jumping output signals of the vision algorithms. To cope with this problem, usually filters are added to the control scheme, such that an adequate motion control signal can be provided for the lower-level controllers.

These filters can be replaced by the OTG algorithm, which can receive pose signals from the image processing algorithm(s) and generate set-points for the underlying control architecture. Besides the advantage mentioned in the previous section (real-time capability of the vision system is not necessarily required), further new options appear:

- The direct specification of motion constraints ($\vec{V}_i^{max}$, $\vec{A}_i^{max}$,$\vec{J}_i^{max}$) becomes possible.
- The interface between the image processing algorithm and the OTG algorithm can either be pose-based ($\vec{P}_i^{trgt}$) only, or pose- and velocity-based ($\vec{V}_i^{trgt}$), such that the vision system can specify velocity vectors in space that are passed through by the OTG module. This has two major advantages:

  1. If the current value of the image error function exceeds a certain value, a trajectory can be automatically generated by the underlying OTG algorithm, such that a smooth and fast motion can be performed.
  2. Vision-based tracking applications that utilize estimation techniques, commonly probabilistic filters, can forward the estimated values by means of velocity vectors in space to the OTG algorithm, such that the interface would be extremely simple, and the resulting motions would be continuous and kinetically time-optimal.

- If the image processing algorithm looses track of its features and is not able anymore to generate adequate command variables for the OTG algorithm, pre-defined fallback parameters can be utilized by the OTG, such that continuous and adequate motions can always be exerted, even if the vision system fails (cf. Chap. 7.2.5, p. 126, and Chap. 7.4, p. 132).

9.3 Relation to High-Level Motion Planning Systems

Another use-case for the OTG algorithm, in which we can achieve novel advances, is based on its relation to higher-level motion planning systems. As already introduced in Chap. 2.4.1 (p. 16), we consider a special field of this area: real-time adaptive motion planning (RAMP, [266]). We assume robots that have to act in a dynamic and/or unknown environment, and that are equipped with sensor systems to react to (unknown) static or dynamic obstacles, events, or abrupt changes of task parameters.

The OTG algorithm can act as an interface for *knots* [266] or *milestones* [280], which are calculated by a higher-level system. These knots or milestones are generated from an overall and global view of the robotic system and its environment. Thereby, it is of course very important to quickly react to unforeseen changes in the environment. All these requirements cannot be provided by the here-presented concept of on-line trajectory generation, but the output values of such higher-level motion planning systems, that is, the knots and milestones, can be ideally used as input values for the OTG algorithm, such that the combination of these systems leads to a very good symbiosis. According to the very abstract three-layer model for robot motion control of Fig. 1.5 (p. 5), the high-level motion planning system is responsible for global planning and sends trajectory parameters (knots, milestones) to the OTG module of the hybrid switched-system controller, which finally generates set-points for the low-level actuator controllers (cf. Fig. 7.1, p. 106). Based on this idea, we can realize robotic systems that can move according to global and task-dependent motion planning and that do not loose the ability to knee-jerkily react to low-level sensor events.

In the following, we explain this idea by means of a concrete, simple, and static example. For illustration, Fig. 9.1 depicts a configuration space obstacle in two-dimensional space. The task of the robot is to move from $\vec{P}_0 = (50, 50)\,mm$ to $\vec{P}_0^{\,trgt} = (700, 300)\,mm$. For this purpose, the high-level motion planning system may calculate H intermediate motion states

$$ {}^{h}\mathbf{M}_0^{trgt} = \left({}^{h}\vec{P}_0^{\,trgt}, {}^{h}\vec{V}_0^{\,trgt}, {}^{h}\vec{A}_0^{\,trgt} \right) \quad \text{with } h \in \{1, \ldots, H\}\ , \qquad (9.2) $$

which have to be passed through by the trajectories generated by the OTG algorithm. H is the number of calculated motion states that correspond to the knots or milestones, that is, the motion states of eqn. (9.2) constitute

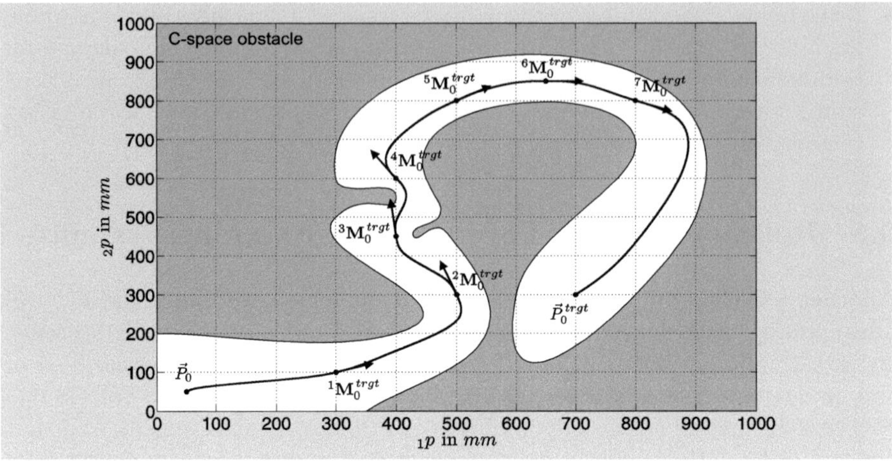

Fig. 9.1 A high-level motion planning system can calculate intermediate motion states ${}^h\mathbf{M}_0^{trgt}$ $\forall\, h \in \{1,\ldots,7\}$ in configuration space, which are passed through by the on-line generated trajectories from $\vec{P}_0$ to $\vec{P}_0^{trgt}$.

the interface to the high-level motion planning system. Of course, the motion constraints ${}^h\mathbf{B}_i$ can also be adapted for the motions in-between two knots. In the case of dynamic environments, all motion states that have not been passed by the robot yet can be furthermore adapted at any future time instant, of course.

The result of Fig. 9.1 was achieved with the Type IV OTG algorithm, that is, ${}^h\vec{A}_0^{trgt} = \vec{0}\ \forall\, h \in \{1,\ldots,7\}$, such that velocity vectors were simply put into the given configuration space:

$$\begin{aligned}
{}^1\vec{P}_0^{trgt} &= (300,100)\,mm & {}^1\vec{V}_0^{trgt} &= (80,30)\,mm/s\\
{}^2\vec{P}_0^{trgt} &= (500,300)\,mm & {}^2\vec{V}_0^{trgt} &= (-30,100)\,mm/s\\
{}^3\vec{P}_0^{trgt} &= (400,450)\,mm & {}^3\vec{V}_0^{trgt} &= (-10,100)\,mm/s\\
{}^4\vec{P}_0^{trgt} &= (400,600)\,mm & {}^4\vec{V}_0^{trgt} &= (-50,80)\,mm/s\\
{}^5\vec{P}_0^{trgt} &= (500,800)\,mm & {}^5\vec{V}_0^{trgt} &= (60,40)\,mm/s\\
{}^6\vec{P}_0^{trgt} &= (650,850)\,mm & {}^6\vec{V}_0^{trgt} &= (120,0)\,mm/s\\
{}^7\vec{P}_0^{trgt} &= (800,800)\,mm & {}^7\vec{V}_0^{trgt} &= (150,-70)\,mm/s\ .
\end{aligned} \tag{9.3}$$

Finally $\vec{P}_0^{trgt} = (700,300)\,mm$ is reached with zero-velocity.

This section introduced how the OTG can act as an interface to high-level motion planning systems. The major symbiotic effect of this use-case is that a robotic system that is guided by a higher-level planning system can obtain the ability of performing immediate reflex motions as a reaction to unforeseen events.

9.4 The Problem of Overshooting

Before we are able to continue with the embedding of dynamics into the proposed on-line trajectory generation framework in the next section, we should discuss a natural problem of this framework as it has been described up to now. Chaps. 4 and 5 (p. 45 and p. 69) show that there is always one, and only one, time-optimal trajectory transferring the system from the current state of motion $\mathbf{M}_i$ to the desired target state of motion $\mathbf{M}_i^{trgt}$.

As long as $\vec{V}_i^{trgt}$, $\vec{A}_i^{trgt}$, and $\vec{J}_i^{trgt}$ (Type IX) equal the zero vector, this is a fully sufficient solution. In the cases of using the OTG algorithm

- as an interface to non-real-time systems/environments (cf. Sec. 9.1)
- as an intermediate layer in visual servo control architectures (cf. Sec. 9.2)
- as an interface to higher-level on-line motion planning concepts (cf. Sec. 9.3), or
- for high-performance motions, in which system dynamics play a significant role,

it is required to specify a set of velocity vectors $\vec{V}_i^{trgt}$ (or even also acceleration and jerk vectors, $\vec{A}_i^{trgt}$ and $\vec{J}_i^{trgt}$) in space, but this fact may lead to undesired trajectories. Such a trajectory (Type IV) and its corresponding path are exemplarily shown for a two-DOF system in Figs. 9.2 and 9.3. The overshooting of DOF 1 is undesired and is caused by the fact that $\vec{V}_i^{trgt}$ (, $\vec{A}_i^{trgt}$, and $\vec{J}_i^{trgt}$) are reached co-instantaneously in $\vec{P}_i^{trgt}$. To prevent this, $\vec{P}_i^{trgt}$, $\vec{V}_i^{trgt}$, $\vec{A}_i^{trgt}$, and $\vec{J}_i^{trgt}$ have to be decoupled, that is, instead of reaching the whole motion state,

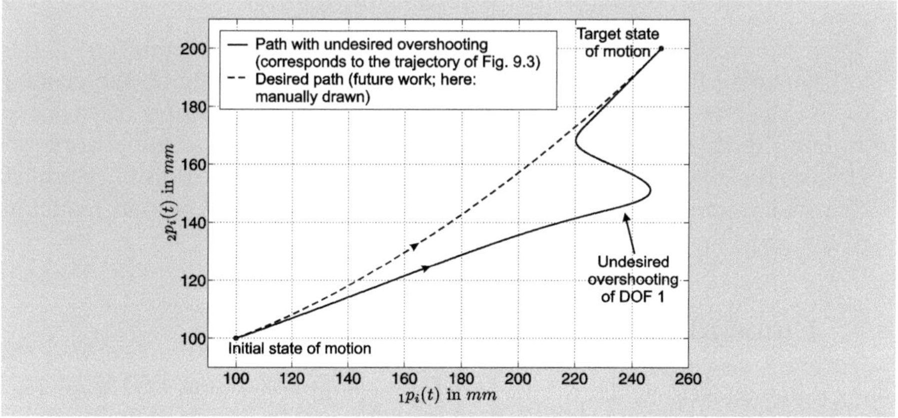

Fig. 9.2 *Path* of a two-DOF Type IV trajectory with undesired overshooting (solid line, cf. Fig. 9.3) and of a possible trajectory without overshooting (dashed line).

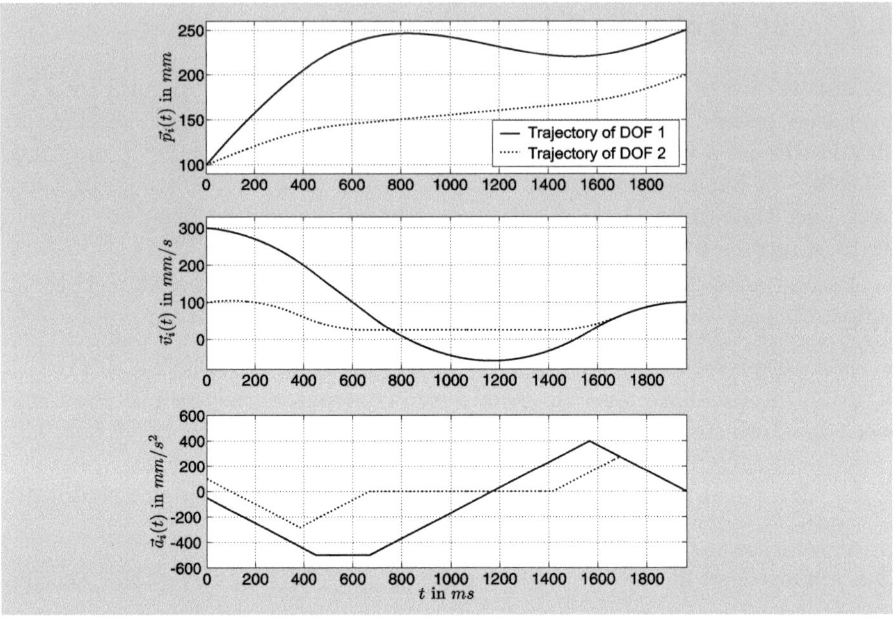

Fig. 9.3 Jerk-limited trajectory generated on-line *from an arbitrary state of motion* and corresponding to the solidly drawn path of Fig. 9.2.

- only $\vec{P}_i^{trgt}$, $\vec{V}_i^{trgt}$, and $\vec{A}_i^{trgt}$ (Type IX, cf. Table 3.1, p. 39),
- only $\vec{P}_i^{trgt}$ and $\vec{V}_i^{trgt}$ (Types V,VII, and IX), or
- only $\vec{P}_i^{trgt}$ (Types II, IV, V, and VII – IX)

are reached, and overshootings, as shown in Figs. 9.2 and 9.3, become prevented.

From the application's point of view, the only possible solution of the algorithms described in Chaps. 4 and 5, must not necessarily be the desired one. To calculate desired solutions that prevent this overshooting, the theory introduced in this monograph has to be extended in a forthcoming work.

Besides the use-cases described in Sections 9.1 – 9.3, this extension — which is clearly a *future work* — can also be applied if a dynamic system model is used for trajectory optimization.

9.5 Embedding of Robot Dynamics

The on-line trajectory algorithm, as it has been introduced in this book, can only generate kinetically time-optimal trajectories (cf. Chap. 3.3, p. 40) as only constant kinematic motion constraints are considered. This is already sufficient for many fields of application, in particular those fields, in which only relatively low robot velocities are exerted. But it is of course a disadvantage of major importance for applications that require high-performance robot motions to utilize trajectories that also consider the system dynamics. This topic was introduced in Chap. 2.4.2 (p. 17), and we can find a huge number of off-line trajectory generation concepts that make use of a dynamic system model. Most approaches have their roots in the concepts of Bobrow, Pfeiffer, and Shin [27, 28, 208, 238, 239].

A dynamic forward model of a mechanical system describes the resulting trajectory, in particular the resulting acceleration progression, in relationship to the applied actuation forces and/or torques. The inverse dynamic model calculates forces/torques if a trajectory (i.e., an acceleration progression) is given. As with trajectory generation, the description of rigid body system dynamics belongs to the very basic levels in the field of robotics. The work of Featherstone [81] was one milestone in the early 1980s. General overviews are given in [82] and [83]. In the following, we do not consider the dynamics of a concrete robotic system, but suggest a generic approach that can be applied to a number of robotic systems.

If we extend the proposed OTG algorithms and consider system dynamics, the OTG input value $\vec{A}_i^{max}$ can no longer be considered as constant. Assuming the actuation forces and/or torques[1] at time T_i are denoted by

$$\vec{F}_i = \left({}_1F_i, \ldots, {}_kF_i, \ldots, {}_KF_i\right) \tag{9.4}$$

and the system-dependent (constant) maximum values of $\vec{F}^{max}$, the forward dynamics can be regarded as a differential equation

$$\ddot{\vec{P}}_i = \vec{f}\left(\vec{P}_i, \dot{\vec{P}}_i, \vec{F}_i\right) \quad \Longleftrightarrow \quad \vec{A}_i = \vec{f}\left(\vec{P}_i, \vec{V}_i, \vec{F}_i\right) \tag{9.5}$$

and in particular

$$\vec{A}_i^{max} = \vec{f}\left(\vec{P}_i, \vec{V}_i, \vec{F}^{max}\right) . \tag{9.6}$$

The latter equation may lead to the naive and simple *but wrong* approach to feed the vector $\vec{A}_i^{max}$ of eqn. (9.6) back as input value for the OTG. This simple idea could only work for constant or continuously increasing values for all elements of $\vec{A}_i^{max}$ (decreasing values would lead to an overshooting of $\vec{P}_i^{trgt}$ due to the reduced deceleration ability).

To incorporate system dynamics, the whole trajectory has to be considered. Respective algorithmic concepts are computationally much more expensive than the proposed OTG algorithms and, hence, cannot be executed at every low-level control cycle T^{cycle}. Because of this fact, we will never achieve time-optimal trajectories with the concept of OTG, but only near-time-optimal ones. The basic idea is to choose a time interval

$$T^{adapt} = \lambda \cdot T^{cycle} \text{ with } \lambda \in \mathbb{N}\backslash\{0\} \; ; \tag{9.7}$$

[1] Prismatic joints exert forces, and rotational joints exert torques.

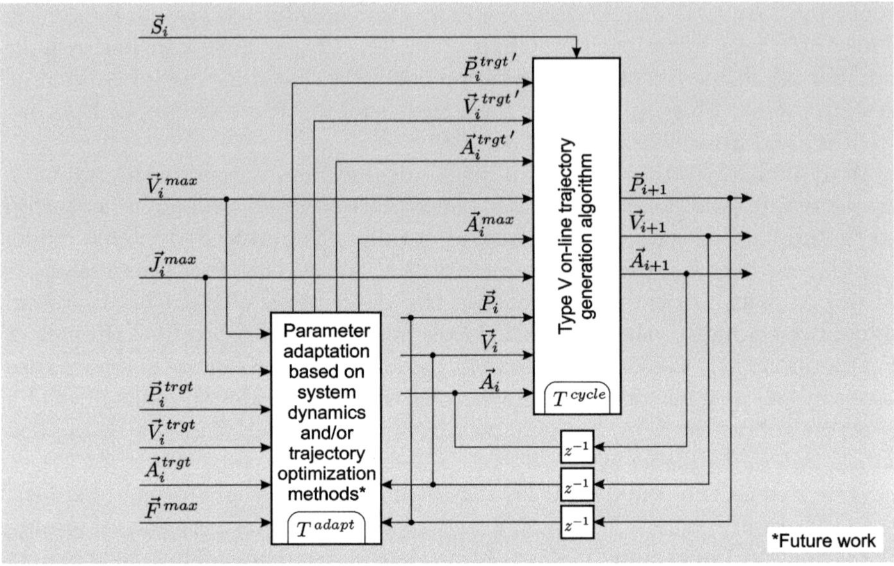

Fig. 9.4 Scheme for Type V OTG with parameter adaptation based on system dynamics and/or trajectory optimization methods (cf. Fig. 3.3, p. 37). As indicated in the two blocks, the trajectory optimization algorithm is executed once per T^{adapt} while the OTG algorithm has a λ times lower cycle time of T^{cycle}.

the trajectory optimization algorithm is executed once per T^{adapt} and adapts the input parameters of the OTG algorithm, as shown in Fig. 9.4. This figure shows only the idea; the contents of the block *Parameter adaptation based on system dynamics and/or trajectory optimization methods* is a subject of future work. Depending on the task and on the chosen trajectory optimization method, the algorithm of this block may adapt the elements of $\vec{A}_i^{max}$, but it is also possible to calculate intermediate states of motion consisting of $\vec{P}_i^{trgt\,\prime}$, $\vec{V}_i^{trgt\,\prime}$, and $\vec{A}_i^{trgt\,\prime}$.

These automatically computed motion states $\mathbf{M}_i{}'$ are used as input values for the actual OTG algorithm, but we have to prevent from undesired overshootings as discussed in the previous section. As a consequence the values of $\mathbf{M}_i{}'$ may be adapted to $\vec{V}_i^{\underline{trgt}}$ and/or $\vec{A}_i^{\underline{trgt}}$ and/or $\vec{J}_i^{\underline{trgt}}$ (cf. Sec. 9.4).

To demonstrate this procedure, Fig. 9.5 shows a simple point-to-point trajectory for a two-DOF mechanism whose maximum acceleration values are adapted depending on the system dynamics. The adaptation process runs with a cycle time of $T^{adapt} = 1\,s$ (this is only an example), and the cycle time of the OTG algorithm is $T^{cycle} = 1\,ms$ ($\lambda = 1000$). In this concrete example, we adapt the maximum acceleration values $\vec{A}_i^{max}$, and within each adaptation step, it must be assured that the applied values of $\vec{A}_i^{max}$ do not exceed their respective maxima. In the example of Fig. 9.5, the following

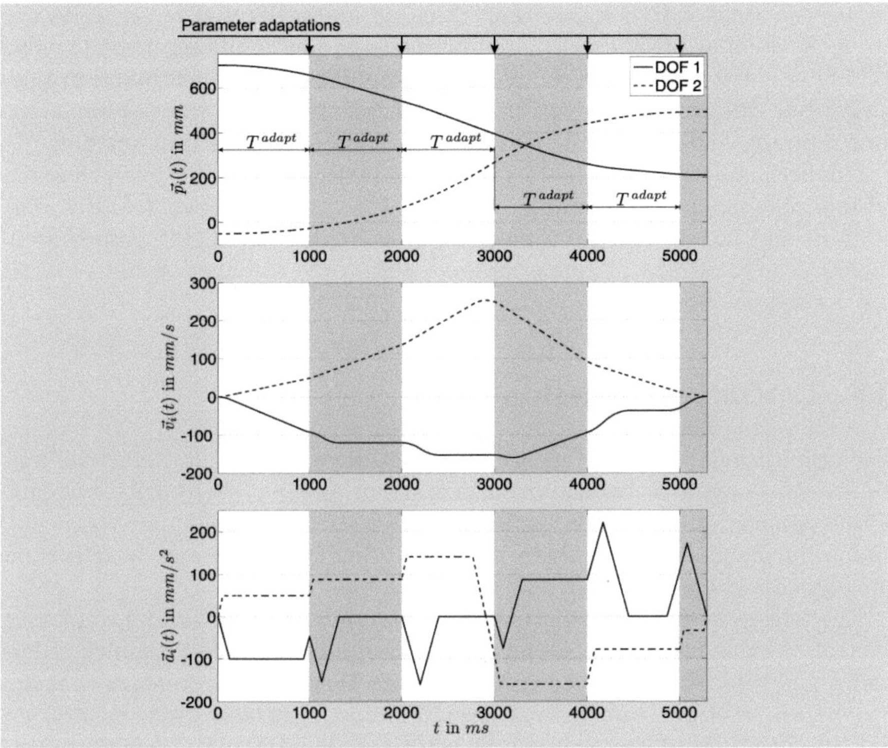

Fig. 9.5 Motion trajectory of a two-DOF system taking into consideration a dynamic model and with one parameter adaptation per second, $T^{adapt} = 1\,s$.

values were obtained in order to bring the system from $\vec{P}_0 = (700, -50)^T\,mm$ to $\vec{P}_i^{trgt} = (180, 490)^T\,mm \;\forall\; i \in \{0, \ldots, 5294\}$:

$$\begin{array}{llll} \vec{A}_0^{max} = (99, 50)^T & mm/s^2 & \vec{A}_{3000}^{max} = (87, 159)^T & mm/s^2 \\ \vec{A}_{1000}^{max} = (212, 88)^T & mm/s^2 & \vec{A}_{4000}^{max} = (223, 78)^T & mm/s^2 \\ \vec{A}_{2000}^{max} = (181, 141)^T & mm/s^2 & \vec{A}_{5000}^{max} = (173, 34)^T & mm/s^2\ . \end{array} \tag{9.8}$$

At $t = 4000\,ms$ and at $t = 5000\,ms$, when the maximum acceleration value for DOF 2 jumps from ${}_2A_{3999}^{max} = 159\,mm/s^2$ to ${}_2A_{4000}^{max} = 78\,mm/s^2$ or from ${}_2A_{4999}^{max} = 78\,mm/s^2$ to ${}_2A_{5000}^{max} = 34\,mm/s^2$, respectively, one can recognize the functionality of OTG Variant B, that is, the system works with motion state values that are outside of the given maximum values. By means of the dashed line in the acceleration diagram of Fig. 9.5, one can recognize that the maximum acceleration value of DOF 2, ${}_2A_i^{max}$, which is derived from ${}_2F^{max}$ (cf. eqn. (9.6)), is the limiting value for the time-optimal execution, taking dynamics into consideration.

This example shows that we are principally able to perform highly dynamic motions while maintaining the capability to instantaneously react to unforeseen events. It is clear, that these on-line generated trajectories are not as optimal as off-line calculated ones. Regarding minimum-time (or also minimum-energy) trajectories, off-line concepts result in much better trajectories, but due to the concept proposed here, we obtain the possibility to interrupt any highly dynamic motion at any unpredictable time instant.

Furthermore, it is possible to generate a trajectory off-line, such that the off-line planner is a submodule of a hybrid switched-system (cf. Fig. 7.1, p. 106), and in an unforeseen moment of switching, the OTG immediately supersedes the off-line planning module. This topic will also be addressed in the discussion about future work in Chap. 10.2 (p. 182).

9.6 Further Applications

The applications that were mentioned up to here focused on industrial manipulation systems or, in the one-dimensional case, on servo drive systems. These applications are often confined to well-structured, safe environments, and well-defined tasks, but there are many other fields that may benefit from the conceptual idea of OTG.

The advent of *service robots* calls for more dynamic approaches to motion control. Service robots are supposed to operate in the direct vicinity of humans, and typically they even cooperate with them in environments that are usually unstructured and/or unknown. The instantaneous utilization of on-line modifications of pre-planned trajectories in case of unforeseen obstacles, for example, the human himself, leads to a great benefit for this area.

As already reviewed in Chap. 2.4.4 (p. 27), the field of *humanoid robotics* commonly has the same requirements as previously mentioned for the field of service robotics. If something unforeseen happens, the robot has to react instantaneously. Due to the reflex motion capability of the OTG algorithm, such robotic reflexes become feasible (cf. Chap. 1.2, p. 6).

The field of *robotic surgery* also calls for such concepts. In this area, safety and redundancy are of major importance, and we need concepts to safely react to unforeseen events or malfunctions. Depending on the situation, a simple quick stop of a robotic surgery system need not necessarily be the correct behavior, because decelerating a robot to zero velocity may be undesired and could hurt a patient. The OTG algorithm would be able to take over control in any situation and could guide the robot to a safe state of motion. Due to its determinism, the algorithm can also be executed on redundant controllers in order to apply this set-point generating instance in a redundant way.

As described in Sec. 9.1, the OTG algorithm can be used as an interface to non-real-time systems. This is an important feature that can be advantageously used in the field of *tele-manipulation.* If sensor events happen on the tele-manipulated manipulator side, switchings as described in Sections 8.4 (p. 145), 8.5 (p. 148), and 8.6 (p. 149) can be performed autonomously and under real-time conditions without instantaneous interaction of the operator, who is connected to the system via a non-real-time and/or delayed communication channel.

The jerk limitation, which can be achieved with the OTG algorithm (except Type I and II), is of immense interest for the control of *parallel kinematic machines*, for example, a HEXA-kinematic [213], because these machines require jerk-limited trajectories in principle [220].

Besides these practical applications, also theoretical considerations are possible. Since the OTG algorithm can compute output values for arbitrary states of motion, there is a chance for it to become relevant for stability analysis methods in the field of control engineering, in particular in the field of hybrid switched-system analysis.

9.7 Migration of Existing Architectures

Although a significant part of this work (Chaps. 3 – 6) is of a theoretical nature, the step for using the OTG algorithm widely in day-to-day practice is only a small one. Although robot manufacturers [3, 60, 79, 123, 153, 186, 189, 194, 250] commonly do not reveal details about their control schemes and architectures, one can suppose that joint control schemes similar to the one below the dashed line of Fig. 7.1 (p. 106) are utilized. Assuming that such an interface to the joint control level, that is, the dashed line in Fig. 7.1, exists, the Type IV OTG algorithm can be directly applied without any further effort. Its interface is *very* simple and consists of one C++ method only[2]:

```
int OTG::GetNextSamplePosition(const OTGDoubleVector &CurrentPosition
                          ,        OTGDoubleVector *NewPosition
                          ,  const OTGDoubleVector &CurrentVelocity
                          ,        OTGDoubleVector *NewVelocity
                          ,  const OTGDoubleVector &CurrentAcceleration
                          ,        OTGDoubleVector *NewAcceleration
                          ,  const OTGDoubleVector &MaxVelocity
                          ,  const OTGDoubleVector &MaxAcceleration
                          ,  const OTGDoubleVector &MaxJerk
                          ,  const OTGDoubleVector &TargetPosition
                          ,  const OTGDoubleVector &TargetVelocity
                          ,  const OTGBoolVector   &SelectionVector);
```

The parameters of this method are self-explanatory (cf. Fig. 3.3, p. 37); `OTGDoubleVector` contains an array of `double` values; `OTGBoolVector` contains an array of Boolean values. The constructor

```
OTG(const unsigned int NoOfDOFs, const double CycleTimeInSeconds);
```

is of a very simple nature as well, such that the application of the OTG algorithm in joint space is absolutely trivial and does not require much effort. The migration to Cartesian space can usually not be generalized and

[2] These code extracts are exact copies of the current OTG implementation.

depends on the existing architecture and kinematic structure. If respective transformation algorithms, as shown in Fig. 7.1, are available, the OTG algorithm can easily be utilized as an open-loop pose controller. This controller could, for example, run in parallel to existing trajectory generators, and if an unforeseen (sensor) event happens, to which the system has to react, the OTG module can supersede the current control submodule.

9.8 Real-Time Verification

The OTG algorithm and its implementation are designed to be real-time capable, that is, we must be able to specify a worst-case execution time of the algorithm. Depending on the type of the OTG algorithm, $(2\alpha - 4\beta)$ decision trees are required (cf. Table 3.1, p. 39). To calculate the worst-case execution time of the algorithm, we can set up trees in the sense of graph theory [68], which have the same structure as the original decision trees. Each node is assigned the execution time of the respective decision or intermediate trajectory segment, and each leaf is assigned the computation time of the corresponding motion profile. By exerting a complete graph search, we can determine the computationally most expensive path through each tree; thus we can specify a worst-case execution time for each single tree. The sum of these times multiplied by the number of DOFs delivers the worst-case execution time of the OTG algorithm.

One problem during the calculation of the worst-case execution time is that the Anderson-Björck-King method (cf. App. A, p. 185) is an iterative one, and its execution time depends on the input values of the algorithm. In order to maintain robustness, the algorithm has been modified, such that the *stop condition step* (cf. App. A) does not depend on the reaching of a threshold ε, but on the number of cycles that are required to achieve a certain relative accuracy with the bisection method, such that the resulting number is independent of the input values.

As a result for the Type IV OTG algorithm, a maximum execution time value of 90 μs was determined (used hardware: AMD Athlon64 3700+ (2.2 GHz, 1024 KB L2 Cache), 2 GB DDR-400, Gigabyte GA-K8NF9 Ultra F5 Mainboard), such that the worst-case execution time for a robotic system with six DOFs is 540 μs on a low-end single-core machine. If arbitrary values are applied to the algorithm, that is, when random numbers $[-1000, 1000]$ are used for the elements of $\vec{W}_i$, the *average* execution time for six DOFs is 135 μs. This difference of factor four is due to the calculation of inoperative time intervals. In most cases, inoperative time intervals do not exist, but for the worst-case calculation, we consider them for all DOFs, of course.

For speed-up, multi-core machines can be used for parallelization of the algorithm, as indicated in Fig. 9.6 (cf. Figs. 5.1, p. 70, and 6.2, p. 102). The $(2\alpha - 4\beta - 1)$ Step 1 decision trees as well as the Step 2 decision tree can be executed in parallel for all K DOFs. Assuming that we have a K-DOF system

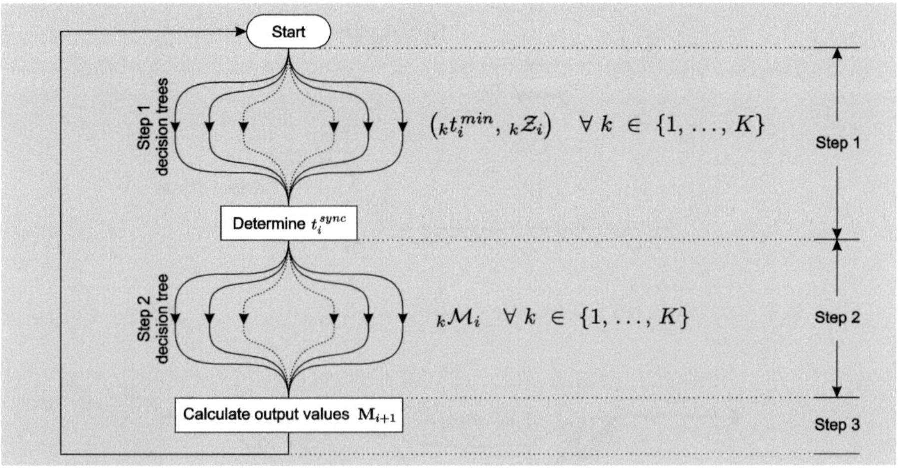

Fig. 9.6 Scheme for the parallelization of the OTG algorithm (cf. Fig. 5.1, p. 70).

and an F-core computing machine, the theoretic performance gain g would be given by

$$g = \frac{K}{\lceil K/F \rceil} \; . \tag{9.9}$$

In practice, of course, we have to consider, that multi-core thread-scheduling and respective context switching is more expensive, such that g is only a theoretic value; the real performance gain depends on the used hardware and the operating system — hence, it has to be measured individually.

9.9 Higher Control Rates

All results of Chap. 8 were achieved with controllers running at a rate of 1 KHz. One can find many publications in the field of force/torque control (cf. Chap. 2.4.3, p. 25) and also in the field of haptics (e.g., [9, 17]) that regard a control rate of 1 KHz as sufficient. Considering the dynamics of the majority of robotic systems, this is correct; in particular, serial kinematic machines *commonly* do not require higher control rates from this point of view.

Nevertheless, the intention of this section is to call attention to higher control rates. One benefit of the OTG concept is that robotic reflexes become enabled, and thereby the reaction time is of fundamental character. The same is, of course, true for human beings: The faster we can react, the more dexterous our reaction behavior will be. But robotic systems commonly feature a much higher level of stiffness, and when a robot, for example, comes into contact with its environment, forces/torques can increase very fast and jerkily. Hence, a minor reason for control rates higher than 1 KHz is surely an

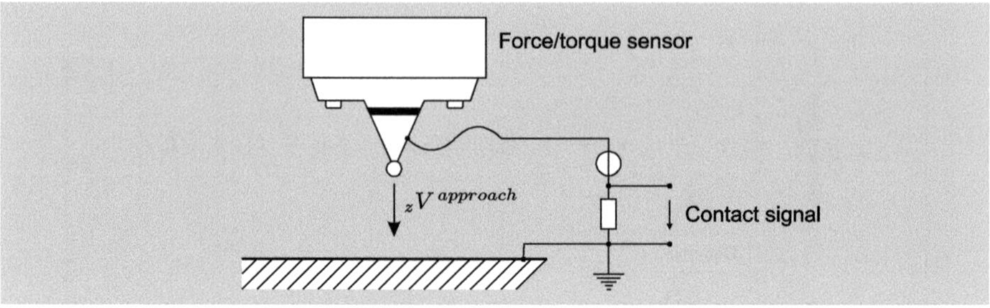

Fig. 9.7 Schematic set-up for the contact experiments of Fig. 9.8 (cf. [85]).

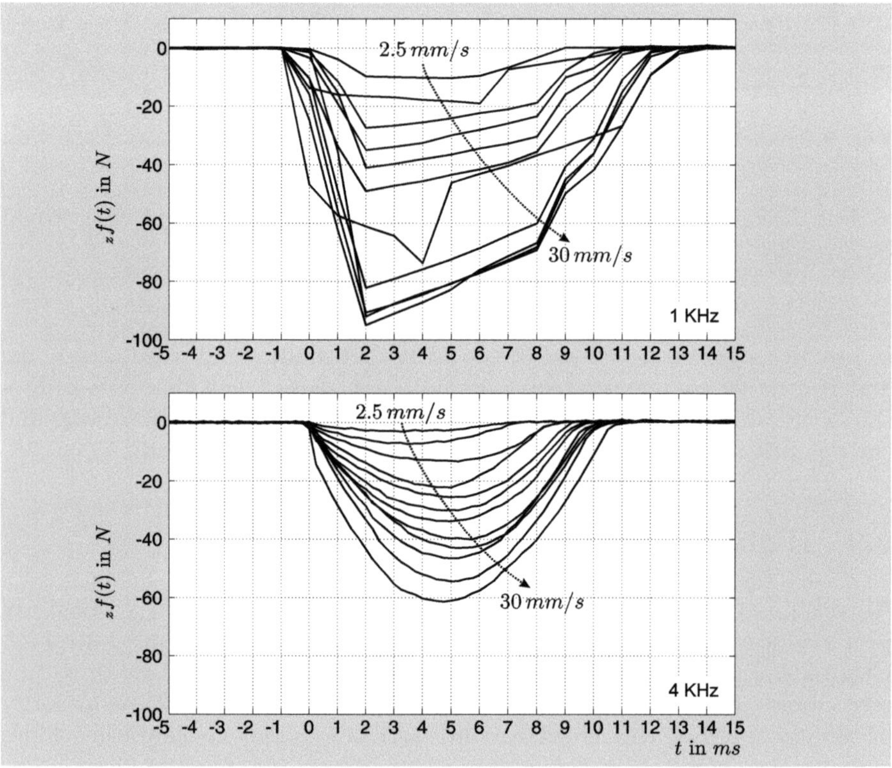

Fig. 9.8 Measured forces during the contact experiment sampled at 1 KHz and at 4 KHz with varying approach velocities $_zV^{approach} \in \{2.5mm/s, 5.0mm/s, \ldots, 30mm/s\}$ (cf. eqn. (9.10), Fig. 9.7, and [85]).

improved control behavior; a much more significant reason for high control rates is the reaction time to sensor events. In particular, stiff robotic systems can achieve the best results with control periods, as short as possible.

To clarify these statements, Fig. 9.7 illustrates the set-up of a very simple experiment.[3] The end-effector of a manutec r2 industrial robot [175] consists of a piece of aluminum that was electrically isolated from the rest of the machine. An electric potential was impressed on this end-effector, and the respective ground potential was impressed on a massive and stiff steel board. The task of the manipulator was to approach the steel board with different velocities,

$$ {}_{z}V^{approach} \in \{2.5\,mm/s, 5.0\,mm/s, \ldots, 30\,mm/s\} , \qquad (9.10) $$

and as soon as the environmental contact was detected by measuring the voltage of the contact signal, the robot was commanded to move away from the steel board as fast as possible. Depending on the rate, at which the system samples the contact signal (cf. Fig. 9.7), the occurring maximum contact forces differ significantly. The top of Fig. 9.8 shows the result for 1 KHz, and the bottom part contains the same curves for 4 KHz.

Due to this fact, issues of software engineering (cf. eqn. (9.9) and Fig. 9.6) become more and more relevant in order to head toward increasing control rates. Published works on real-time software engineering in the field of robotics are, for example, [19, 86, 197, 219, 225].

9.10 Aspects on the Development of Decision Trees

In Remark 5.2 (p. 82), we already started the discussion about the question, how to develop decision trees for the purpose of on-line trajectory generation. Decision trees constitute the heart of all OTG algorithms, because the trees have to guarantee, that eqn. (4.13) (p. 48)

$$ \bigcup_{r=1}^{R} {}^{r}\mathcal{D}_{Step1} \equiv \mathbb{R}^{\alpha} , $$

eqn. (5.18) (p. 78)

$$ \bigcup_{s=1}^{S} {}^{s}\mathcal{D}_{Step2} = \left(\mathbb{R}^{\alpha+1}\right) \setminus \mathcal{H} , $$

and eqn. (5.19) (p. 78)

$$ \mathcal{S} = \bigcup_{s=1}^{S} \left\{ {}^{s}\mathcal{D}_{Step2} \cap {}^{u}\mathcal{D}_{Step2} \ \forall\, u \in \{1, \ldots, S\} \,\middle|\, u \neq s \right\} $$

hold in practice, and that all four criteria, which were introduced in Chap. 3.3 (p. 40), are fulfilled. These issues can only be solved and guaranteed manually,

[3] This experiment was already presented and discussed extensively in the thesis of Finkemeyer ([85], p. 120), and this is only a brief recap of the same subject.

that is, the developer of a decision tree (cf. Figs. 4.4 (p. 52), 4.8 (p. 62), 5.6 (p. 83), 5.7 (p. 85), and 5.8 (p. 86)) has to be very careful in order not to overlook any case. Due to their complexness, one could get the idea to search for a method that develops these trees automatically, but this is an impassable meander, as will be explained in the following.

When starting to develop a decision tree, the motion profile sets $\mathcal{P}_{Step1}$ and $\mathcal{P}_{Step2}$ are unknown. These two sets have to fit exactly to each other, such that the three equations mentioned above hold. The design of these sets can only take place during the development of the $(2\alpha - 4\beta)$ (cf. Table 3.1, p. 39) decision trees, because we cannot know how these profiles look like, until we know all possible cases. Summarizing this paragraph briefly:

- If we knew the motion profile sets $\mathcal{P}_{Step1}$ and $\mathcal{P}_{Step2}$ (and thus all input domains ${}^{r}\mathcal{D}_{Step1}\ \forall\ r \in \{1,\ldots,R\}$ and ${}^{s}\mathcal{D}_{Step2}\ \forall\ s \in \{1,\ldots,S\}$) for a concrete type of OTG, it *might* be possible to generate the decision trees automatically.
- If we knew all $(2\alpha - 4\beta)$ decision trees for a concrete type of OTG, it would be possible to determine the motion profile sets $\mathcal{P}_{Step1}$ and $\mathcal{P}_{Step2}$.

But since we neither know $\mathcal{P}_{Step1}$ or $\mathcal{P}_{Step2}$ of a concrete type nor the decision trees beforehand, it cannot be possible to generate an OTG algorithm automatically.

Furthermore, it is not possible to formulate general laws for the tree development, because every single decision has its own purpose. Of course, there are decisions that work very similarly, but there is always a nuance that is different. One property that can be influenced by the developer is the size of the trees. It depends on the order of the decisions, and there is always one order that leads to a minimum number of decisions. The trees, as they have been proposed for the Type I OTG algorithm in [146], are relatively large, because the order is non-optimal.

9.11 Completeness Analysis of Decision Trees

All decision trees are generated manually, and we have to wonder how we can ensure that all $(2\alpha - 4\beta)$ trees are complete, such that the eqns. (4.13), (5.18), and (5.19) hold. To verify the completeness of all trees, we assign a Boolean variable to each edge.[4]

Initially, a zero is assigned to all edges of all trees. Then, α randomly generated numbers are repetitively applied to the algorithm, and if an edge is passed by the algorithm, its value is set to one. This procedure is repeated

[4] The decision trees of Figs. 4.4, 4.8, 5.6, 5.7, and 5.8 are presented in a compressed way, such that they can be illustrated in this paper format. Therefore, single nodes (and edges) are used multiple times, and the trees look like graphs, although they are trees, if drawn in an exhaustive way. What is meant by an *edge* in the context of this section is an edge of the exhaustively drawn decision tree.

until a value of one has been assigned to *all* edges. During this procedure, no error must occur, that is, the systems of equations for the selected motion profiles have to be solvable in any case. If there is only one case in which the systems of equations are not solvable, one of the two following errors can occur:

- There is an error in one or more of the $(2\,\alpha - 4\,\beta)$ trees.
- The set $\mathcal{P}_{Step1}$ and/or $\mathcal{P}_{Step2}$ is incomplete/erroneous.

Already in the case of the Type IV OTG algorithm, it is a voluminous task and a big effort to perform this evaluation. Only if all edges of all trees have been passed at least once, and if all involved systems of equations are solvable, the completeness of the decision trees is indicated.

Remark 9.1. *After the Type IV implementation successfully passed this test, random numbers were again repetitively applied to the algorithm in order to obtain an additional indication that the realized Type IV OTG algorithm is complete and works numerically robustly. After more than 30 billion error-free cycles, the author aborted the test.*

9.12 OTG Types V – IX

As already mentioned during the classification of different types of OTG (Chap. 3.4, p. 42), the highest type of OTG algorithm that has been developed by the author is the Type IV OTG algorithm. The development of further types is, of course, possible in the same manner as described for Type IV, but the development effort increases, since the dimensions of the input and/or the output space increase according to Table 3.1 (p. 39).

If we were able to specify target acceleration vectors $\vec{A}_i^{trgt}$ (Type V), it would become possible to perform motion transitions from one state space into another state space with a priori specified transitions:

Example 9.1. *Fig. 9.9 shows the two-dimensional path of a motion that is first controlled w.r.t. Cartesian coordinates (dotted line), and from that instant onward, after ${}^{Cart}\mathbf{M}_i^{trgt}$ has been reached, the motion is controlled w.r.t. cylindrical coordinates (dashed line). This transition from Cartesian coordinates to cylindrical coordinates enables the robot to continue its motion exactly along a circular path from the moment of switching on, because we could specify a target acceleration vector ${}^{Cart}\vec{A}_i^{trgt}$. A similar example for Type IV OTG was already given in Chap. 8.5 (p. 148, Figs. 8.10, 8.11, and 8.12).*

An additional advantage of the Type V OTG algorithm, whose development is a part of a future work, can be exploited if a dynamic model is used for the on-line generation of robot motion trajectories. This feature was already discussed in Chap. 9.5 (p. 164).

The development of the Type VI and higher types of the OTG algorithm will be more demanding than the development of the Type V OTG algorithm,

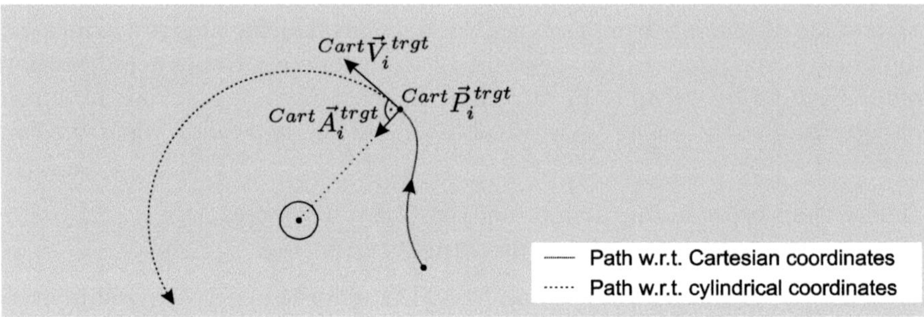

Fig. 9.9 Transition from Cartesian coordinates to cylindrical coordinates via specification of a respective target state of motion ${}^{Cart}\mathbf{M}_i^{trgt}$. Such kinds of transitions can be realized with the Type V OTG algorithm.

because the diversity of motion profiles, that is, jerk profiles, greatly increases. The number of subspaces into which the input domain is subdivided (cf. eqn. (4.13), p. 48) increases significantly as does the number of subspaces for the input domain of Step 2 of the algorithm (cf. eqn. (5.18), p. 78). The advantage of the OTG algorithms of Types VI – IX over the Types III – V OTG algorithms is, that the smoothness of the trajectory increases again, and the effect of avoiding excitations of higher-order dynamics in motion systems by using appropriate kinematic motion constraint values becomes reinforced. As a result, very agile tasks can be realized, for example, transporting a glass completely filled with water from an initial pose to a target pose with a minimum loss of water.

9.13 Further Variants

A drawback of the class of OTG algorithms proposed in Chaps. 3 – 7 is, that the same boundary values V_i^{max}, A_i^{max}, J_i^{max}, and D_i^{max} are used in positive and in negative direction. Depending on the robot and on the control architecture, it can be reasonable to use different bounds for the positive and the negative direction, such that eqn. (3.17) (p. 38) becomes extended to

$$\mathbf{M}_{i+1} = f(\mathbf{W}_i), \text{ where } f:\ \mathbb{R}^{(\alpha+\beta)K} \times \mathbb{B}^K \longrightarrow \mathbb{R}^{\beta K}. \qquad (9.11)$$

In contrast to the realization of higher OTG types, the implementation and integration of this extension can be done without the development of further decision trees. The structure of the trees remains unchanged, and only the decisions and leaves of the trees have to be adapted to the $(\beta\, K)$ further input variables.

If the OTG algorithms were applied in the field of mobile robotics, it would make sense to use different motion constraints for the positive and the negative direction. A car, for example, usually possesses much more deceleration than acceleration power, and, usually, a car can drive much faster in its forward direction than in its backward direction.

9.14 On the Elegance of Natural Laws

Natural laws commonly hold some kind of elegance or beauty [80]. The class of algorithm presented in this book is, of course, not a natural law, but the presented algorithms of this work also feature a kind of elegance.

The development of the Type IV decision trees, for example, takes place in eight and nine dimensional spaces, and the author often felt like a little three-dimensional being searching for something unknown in an eight-dimensional and a nine-dimensional space. It was rarely clear, which way to go in these spaces, how decisions should look, and whether the basic idea of on-line trajectory generation would work at all. But after a while, you go deeper and deeper into these spaces and you discover a kind of beauty, a kind of elegance. Unfortunately — and this is the reason for writing this section — it was not possible to express this kind of elegance in this monograph.

Equations (4.13), (5.18), and (5.19) are the basis for this statement. As indicated in Remark 5.2 (p. 82), the trees of a single type of OTG algorithm are, even in a minimized manner, complex and consist of hundreds of individual decisions. But in the end, after having used several square meters of paper for writing down the decision trees, everything fits *exactly* together — like a wonderful mathematical puzzle.

Chapter 10
Summary, Future Work, and Conclusion

This final chapter gives brief summaries of each single chapter of this monograph, presents ideas for future work, and finally draws conclusions of this work.

10.1 Summary

Literature Survey and Motivation

Besides on-line and off-line path planning and trajectory generation methods, Chap. 2 surveys the fields of robot motion control and hybrid switched-system control. Additionally, force/torque control and visual servo control concepts are discussed as the most common kinds of sensor-guided robot motion control. This survey concludes with the motivation for this work:

"The majority of the surveyed concepts for off-line and also on-line motion generation produce a motion along a specified path. But, is this a good approach? — For purely position/pose and/or trajectory-following controlled motions: Sure and without restriction of any kind! But: When we execute sensor-guided motions, for example, by force/torque or by visual servo control, we do not have a predefined path anyway, because the robot motion directly depends on the sensor signal. We have to dismiss the path during sensor-based motion control! As soon as we embed sensor-guided or sensor-guarded motions, there is no predefined path anymore. In particular, we have to say good-bye to trajectory planning and reference trajectories along previously specified paths. There is no path that can be exactly followed, because everything may depend on sensors whose signals cannot be foreseen."

Introduction of Different Types of OTG Algorithms

Nine different types of OTG algorithms, Types I – IX, have been introduced in Chap. 3. All types generate kinetically time-optimal and time-synchronized

T. Kröger: On-Line Trajectory Generation in Robotic Systems, STAR 58, pp. 179–184.
springerlink.com

trajectories for mechanical systems with one or more DOFs. Depending on the type, the input and output values of the algorithms are specified. We distinguish between types of on-line trajectory generation algorithms that feature acceleration limitation, jerk limitation, or the limitation of the derivative of jerk. Furthermore, different target motion state parameters can be specified depending on the algorithm type. The simplest cases only allow the specification of a target pose/position, and for more complex types, target velocity vectors, target acceleration vectors, and/or target jerk vectors may be specified. The highest type that has been realized by the author is the Type IV OTG algorithm: This algorithm generates jerk-limited trajectories and allows the specification of a target velocity vector.

On-Line Trajectory Generation for One-DOF Systems

The algorithm for multi-DOF systems consists of three steps, and a part of the first step can also be applied to systems with one single DOF. In a one-DOF system, the algorithm time-optimally transfers the system from an arbitrary initial state of motion to a specified target state of motion. After a general description of the concept, the concrete solution for the Type IV OTG algorithm is presented in detail. The basic idea is that a *finite number of motion profiles* exists, of which one profile leads to the time-optimal solution. This profile can be selected by a decision tree, and, subsequently, a corresponding system of equations can be set up and solved. The solution contains all required trajectory parameters, such that the output values, that is, the state of motion for the current control cycle, can be calculated. During the solution of these equations, numerical problems appear, such that a procedure that uses a combination of the Anderson-Björck-King method and the simple bisection method was developed. This procedure ensures the robustness of the algorithm, and deterministic solutions for arbitrary input values can be calculated.

On-Line Trajectory Generation for Systems with Multiple DOFs

The key part of this work is an algorithm for the on-line generation of trajectories for robotic systems with multiple DOFs. This algorithm consists of three steps:

1. Calculation of the minimum synchronization time.
2. Synchronization of all DOFs to the synchronization time of Step 1 and calculation of all trajectory parameters.
3. Calculation of all output values for the current control cycle based on the attained parameters of Step 2.

The proposed algorithm for one-DOF systems is used to calculate the minimum possible execution times for all DOFs. The minimum synchronization time cannot be lower than the maximum time value of all previously calculated times. Depending on the type of OTG, up to three inoperative time intervals may be present for each DOF. These time intervals have to be calculated in the first step of the algorithm, such that we can guarantee, that all DOFs can reach their target state of motion exactly at the instant specified by the synchronization time.

Step 2 of the algorithm calculates all trajectory parameters for all DOFs, such that they reach their target states of motion synchronously, and Step 3 calculates the output values of the current control cycle, which are subsequently used as command variables for lower-level control.

This procedure enables the generation of time-synchronized trajectories in multiple DOFs. Chapter 6 extended this algorithm to be applicable for phase-synchronized trajectories, that is, straight-line trajectories in multi-dimensional space.

Hybrid Switched-System Control

The major advancement that can be achieved with this new algorithm regards sensor-based robot motion control. In Chapter 7, a hybrid switched-control system was suggested that consists of multiple sensors as well as open- and closed-loop controllers, among which the system can switch discretely. How control signals are chosen, is specified by *Manipulation Primitives*, which constitute the interface to higher-level applications. This programming paradigm is formally defined in order to establish the possibility of executing sensor-guided *and* sensor-guarded robot motion commands simultaneously and in a very open way, such that any kind and any number of sensors can be principally addressed. Due to the OTG algorithm, the system is able to switch between different reference coordinate systems at arbitrary (sensor-dependent) instants. Moreover, switchings from task space control to actuator space control (or vice versa) become feasible, such that a safe and robust control system can be realized, since the OTG algorithm is able to take over control from arbitrary states of motion. This feature is very practical and of high relevance because the on-line trajectory generation algorithm can also take over control and/or supersede other controllers if a sensor unexpectedly fails.

Practical Results

After Chaps. 3 – 7 introduced theoretical concepts of on-line trajectory generation, results were presented in Chap. 8. These experiments contain:

Handling arbitrary states of motion
A system with multiple DOFs is transferred from an arbitrary state of motion to a desired target state of motion.

Instantaneous reaction to unforeseen (sensor) events
A simple two-DOF robot executes a motion; an obstacle is unexpectedly detected during the motion, and the OTG algorithm instantaneously generates a time-optimal trajectory to prevent collision and to bypass the obstacle.

Homothetic trajectories
During the execution of a straight-line trajectory, a sensor event happens, and the OTG algorithm immediately generates a new straight-line trajectory.

Unforeseen switchings of reference coordinate systems
A motion that is executed w.r.t. a reference coordinate system, commonly the Task Frame, becomes interrupted by a sensor event, and the control system switches to another reference frame from one control cycle to another, such that the resulting motion remains continuous.

Unforeseen switchings of control spaces
By means of a plain r-φ-manipulator, it is shown how the control system can switch from Cartesian space control to joint space control at an arbitrary instant and in an arbitrary state of motion.

Hybrid switched-system control of a six-DOF industrial manipulator
The last experimental result shown in this book considers a Stäubli RX60 industrial manipulator, which executes a force-controlled motion in some task space DOFs, and at an arbitrary instant, the control system switches to joint space control, and the OTG algorithm takes over control.

Relation to high-level motion planning systems
The proposed concept is designed to act as an open-loop pose control unit at a control level between low-level actuator control and high-level motion planning. An example is shown, in which the OTG algorithm receives a set of velocity vectors in Cartesian space from a kind of high-level system and generates command variables for low-level control.

Embedding of robot dynamics
The proposed algorithm is only capable to generate kinetically time-optimal trajectories, that is, a dynamic system model is not considered. The embedding of dynamics belongs to the subjects of future work (cf. Sec. 10.2). A very simple example is shown in which a maximum acceleration curve is used to generate the trajectory with the proposed OTG algorithm.

10.2 Limitations and Future Work

The proposed concept for the on-line generation of robot motion trajectories is of a very basic nature, and it is an interesting question, whether the suggested algorithms will make their way into day-to-day practice in the course of time. Besides the potential and the advantages that were presented in Sec. 10.1, there are also limitations and open questions.

One very important question that was addressed in Chap. 9.5 is the consideration of robot dynamics, because the generated trajectories are only kinetically time-optimal, which is a significant drawback as only constant kinematic motion constraints are considered. Fig. 9.4 (p. 166) suggests a scheme that generally opens the door for the embedding of a dynamic robot model, but the block *Parameter adaptation based on system dynamics and/or trajectory optimization methods* still is a part of a future work. Only if we are able, as in classical off-line trajectory generation concepts, to incorporate the dynamic behavior of a robot, high-performance motions can be safely exerted.

Another open point is the development of higher types of OTG algorithms. The author was able to develop the Type IV algorithm and described the fundamental concept for all types of on-line trajectory generators. But, as discussed in Chap. 9.13 (p. 176), the Types V and VI would open the door to further features. Type V, for example, is very relevant for the embedding of system dynamics, and Type VI would enable trajectories that feature greater smoothness.

Chap. 7.4 (p. 132) discussed the issue of stability in hybrid switched-control systems using the OTG algorithm, that is, an open-loop pose controller, as one control submodule. The monograph merely contains a derivation of a line of arguments. A formal stability proof is beyond the scope of this work and also part of the future work regarding theoretical hybrid switched-system analysis.

10.3 Conclusion

This work introduced a new concept for motion generation in robotic systems during runtime. The presented on-line trajectory generation algorithm is executed in parallel to low-level motion controllers, such that systems using it are able to react *instantaneously* to unforeseen (sensor) events. In particular, the algorithm closes a significant gap: Switching from sensor-guided machine motions to trajectory-following motions becomes possible at any time and in any state of motion. As a consequence, (multi-)sensor integration becomes greatly simplified, and robot motion control systems are enabled to execute trajectory-following motions, sensor-guided motions, and *sensor-guarded motions*. The proposed on-line trajectory generation algorithm acts as an open-loop pose controller and can take over control at any time instant, such that safe and continuous motions can be guaranteed — even if sensors fail.

Building the bridge to the introductory chapter, in which a brief comparison to the neurophysiological system of human beings was formulated, the proposed algorithm enables robotic systems to perform a kind of *robotic reflex*. This feature enables new fundamental possibilities for robot motion control and programming, and it opens the gate to a new field of sensor-based robotic applications. The concept is very basic and self-contained, such that it is applicable in almost all areas of robotics and many areas of control engineering. Due to its extremely simple interface, the algorithm can be directly used in real-world applications; it works robustly and is real-time capable.

The proposed algorithm was developed with the aim to advance motion control systems of many robotic applications in various fields, such as service robotics, manipulation control systems, mobile robotics and manipulation, or robotic surgery — in short: all fields in which sensor integration plays a fundamental role. The algorithm can be regarded as an intermediate control layer that is one element of the important bridge between low-level robot motion control and higher-level (sensor-based) motion planning.

The basic idea of this work is of simple nature, and the mathematical fundamentals used in this work are of a very basic nature, too. During the entire development of the theory on on-line trajectory generation and during the writing of this book, the author's intention was always to keep everything simple. In order to connect the final sentence of this work with the preface, he would like to close his work with a famous quote from Leonardo da Vinci: *"Simplicity is the ultimate sophistication."*

Appendix A
The Modified Anderson-Björck-King Method

As presented in Chap. 4.2.1 (p. 50), the method of Anderson and Björck [12] with the improvement of King [131] is applied in this work. In order to improve the robustness (at the expense of efficiency), this method was combined with the bisection method as will be described in this appendix.

The Anderson-Björck-King method is a variant of the *regula falsi* and an improvement of the *Pegasus method* [70, 71]. We assume the function f to be continuous in the closed interval $[a,b]$ and that $f(a)f(b) < 0$. Hence, we know that there is at least one root of odd order within the interval of $]a,b[$. The Anderson-Björck-King method calculates one of these zeros by a repeated minimization of the root inclusion interval; it always converges if initially $f(a)f(b) < 0$.

The regarded functions f in the context of OTG can be of very different characteristics, which depend on the input values $\mathbf{W}_i$ at an instant T_i. Although we can always determine an interval $[a,b]$, from which we know that it contains the desired root x_0, it may happen that a (in contrast to the interval width) relatively large part of the function lies in parallel to the abscissa. Such functions lead to efficiency problems when applying the Anderson-Björck-King method, and in the worst case, the method will converge very slowly, such that we are not able to specify a worst-case execution time of the algorithm. To bypass this problem, to improve the robustness, and to be able to specify a maximum number of required iterations in order to be real-time capable, we combine the Anderson-Björck-King method with the simple bisection method. The first three steps of the following algorithm belong to the bisection method and all following ones to the Anderson-Björck-King method.

Given: $f : [a,b] \stackrel{cont}{\longrightarrow} \mathbb{R}$, $f(a)f(b) < 0$, and an error threshold $\varepsilon > 0$ with $\varepsilon \in \mathbb{R}$.

Task: Find an approximation of one root $x_0 \in]a,b[$, such that the difference between the last two iterated values is smaller than ε.

Start: Take $x_1 := a$, $x_2 := b$ as initial values and compute $f_1 := f(x_1)$, $f_2 := f(x_2)$.

The iteration consists of the following seven steps:

1. Bisection Step

$$x_3 := \frac{x_1 + x_2}{2} \tag{A.1}$$

2. Calculation Step I

$$f_3 := f(x_3) \tag{A.2}$$

If $f_3 = 0$, the method ends with $x_0 = x_3$.

3. Interval Determination Step I
Determine a new inclusion interval: If $f_3 f_2 < 0$, x_0 lies between x_2 and x_3, and we set:

$$x_1 := x_2 \tag{A.3}$$
$$x_2 := x_3 \tag{A.4}$$
$$f_1 := f_2 \tag{A.5}$$
$$f_2 := f_3 \,. \tag{A.6}$$

If $f_3 f_2 > 0$ (x_0 lies between x_1 and x_3), we set

$$x_2 := x_3 \tag{A.7}$$
$$f_2 := f_3 \,. \tag{A.8}$$

4. Secant Step
Compute the slope of the connecting line from (x_1, f_1) to (x_2, f_2):

$$s_{12} = \frac{f_1 - f_2}{x_1 - x_2} \tag{A.9}$$

and set

$$x_3 := x_2 - \frac{f_2}{s_{12}} \,. \tag{A.10}$$

5. Calculation Step II

$$f_3 := f(x_3) \tag{A.11}$$

If $f_3 = 0$, the method ends with $x_0 = x_3$.

6. Interval Determination Step II
Determine a new inclusion interval: If $f_3 f_2 < 0$, x_0 lies between x_2 and x_3, and we set:

$$x_1 := x_2 \tag{A.12}$$
$$x_2 := x_3 \tag{A.13}$$
$$f_1 := f_2 \tag{A.14}$$
$$f_2 := f_3 \; . \tag{A.15}$$

If $f_3 f_2 > 0$ (x_0 lies between x_1 and x_3), we assign a new functional value at x_1:

If $1 - \frac{f_3}{f_2} \leq 0$, take

$$g := 0.5 \tag{A.16}$$

otherwise use

$$g := 1 - \frac{f_3}{f_2} \; . \tag{A.17}$$

Then set

$$x_2 := x_3 \tag{A.18}$$
$$f_1 := g f_1 \tag{A.19}$$
$$f_2 := f_3 \; . \tag{A.20}$$

7. Stop Condition Step

If $|x_2 - x_1| \leq \varepsilon$, stop the iteration.[1] If $|f_2| \leq |f_1|$ then set

$$x_0 := x_2 \tag{A.21}$$

otherwise

$$x_0 := x_1 \; . \tag{A.22}$$

If $|x_2 - x_1| > \varepsilon$, the iteration is continued with Step 1 with the new values x_1, x_2, f_1, and f_2 from Step 6.

Further details and an explanation with geometric interpretation of the Anderson-Björck-King method can be found in the book of Engeln-Müllges and Uhlig [73] and of course in [12, 131]. The original Anderson-Björck-King method is the most efficient numerical inclusion method regarding the efficiency index of Traub [264].

[1] For practical implementation, an error threshold of $\varepsilon = 10^{-12}$ was applied.

Appendix B
Details on the *PosTriNegTri* Acceleration Profile (Step 1)

B.1 Position-Error Function

Written form of the position-error function for the Step 1 *PosTriNegTri* acceleration profile:

$$
{}^{PosTriNegTri}P_i^{err1}\left(a_i^{peak1}\right) = \frac{1}{12\,(J_i^{max})^2}\left(4\,(A_i)^3 - 12\,(A_i)\,(J_i^{max})\,V_i - 3\,(A_i)^2\right.
$$

$$
\left(4\left(a_i^{peak1}\right) + \sqrt{-2\,(A_i)^2 + 4\left(a_i^{peak1}\right)^2 + 4\,(J_i^{max})\,(V_i - V_i^{trgt})}\right)
$$

$$
+6\left(2\left(a_i^{peak1}\right)^3 + 4\left(a_i^{peak1}\right)(J_i^{max})\,V_i + (J_i^{max})\left(2\,(J_i^{max})\,(P_i - P_i^{trgt}) + (V_i + V_i^{trgt})\right.\right.
$$

$$
\left.\sqrt{-2\,(A_i)^2 + 4\left(a_i^{peak1}\right)^2 + 4\,(J_i^{max})\,(V_i - V_i^{trgt})}\right)
$$

$$
\left.\left.+\left(a_i^{peak1}\right)^2\sqrt{-2\,(A_i)^2 + 4\left(\left(a_i^{peak1}\right)^2 + (J_i^{max})\,(V_i - V_i^{trgt})\right)}\right)\right).
$$

B.2 Derivative of the Position-Error Function

Written form of the derivative of the position-error function for the *PosTriNegTri* acceleration profile:

$$
{}^{PosTriNegTri}P_i^{err1\prime}\left(a_i^{peak1}\right) = \left(6\left(a_i^{peak1}\right)^3 + 6\left(a_i^{peak1}\right)(J_i^{max})\,V_i - 2\left(a_i^{peak1}\right)\right.
$$

$$
(J_i^{max})\,V_i^{trgt} + 3\left(a_i^{peak1}\right)^2\sqrt{-2\,(A_i)^2 + 4\left(a_i^{peak1}\right)^2 + 4\,(J_i^{max})\,(V_i - V_i^{trgt})}
$$

$$
+2\,(J_i^{max})\,V_i\sqrt{-2\,(A_i)^2 + 4\left(a_i^{peak1}\right)^2 + 4\,(J_i^{max})\,(V_i - V_i^{trgt})}
$$

$$+ (A_i)^2 \left(-3 \left(a_i^{peak1} \right) - \sqrt{-2 (A_i)^2 + 4 \left(a_i^{peak1} \right)^2 + 4 (J_i^{max}) (V_i - V_i^{trgt})} \right) \Bigg)$$

$$/ \left((J_i^{max})^2 \sqrt{-2 (A_i)^2 + 4 \left(a_i^{peak1} \right)^2 + 4 (J_i^{max}) (V_i - V_i^{trgt})} \right) .$$

B.3 Setting up the Parameters of $\mathcal{M}_i$

This appendix section details the parameter calculation of $\mathcal{M}_i$ of Chap. 4.2.1 (p. 58). We first repeat the system of eqns. (4.21) – (4.32), which are required as a base for the following calculations. The unknown variables are t_i^{min}, ${}^2t_i$, ${}^3t_i$, ${}^4t_i$, ${}^2v_i$, ${}^3v_i$, ${}^4v_i$, ${}^2p_i$, ${}^3p_i$, ${}^4p_i$, a_i^{peak1}, and a_i^{peak2}.

$${}^2t_i - T_i = \frac{\left(a^{peak1} - A_i\right)}{J_i^{max}} \tag{B.1}$$

$${}^3t_i - {}^2t_i = \frac{a^{peak1}}{J_i^{max}} \tag{B.2}$$

$${}^4t_i - {}^3t_i = -\frac{a^{peak2}}{J_i^{max}} \tag{B.3}$$

$$t_i^{min} - {}^4t_i = -\frac{a^{peak2}}{J_i^{max}} \tag{B.4}$$

$${}^2v_i - V_i = \frac{1}{2} \left({}^2t_i - T_i\right) \left(A_i + a^{peak1}\right) \tag{B.5}$$

$${}^3v_i - {}^2v_i = \frac{1}{2} \left({}^3t_i - {}^2t_i\right) a^{peak1} \tag{B.6}$$

$${}^4v_i - {}^3v_i = \frac{1}{2} \left({}^4t_i - {}^3t_i\right) a^{peak2} \tag{B.7}$$

$$V_i^{trgt} - {}^4v_i = \frac{1}{2} \left(t_i^{min} - {}^4t_i\right) a^{peak2} \tag{B.8}$$

$$\begin{aligned} {}^2p_i - P_i = {} & V_i \left({}^2t_i - T_i\right) + \frac{1}{2} A_i \left({}^2t_i - T_i\right)^2 \\ & + \frac{1}{6} J_i^{max} \left({}^2t_i - T_i\right)^3 \end{aligned} \tag{B.9}$$

$$\begin{aligned} {}^3p_i - {}^2p_i = {} & {}^2v_i \left({}^3t_i - {}^2t_i\right) + \frac{1}{2} a^{peak1} \left({}^3t_i - {}^2t_i\right)^2 \\ & - \frac{1}{6} J_i^{max} \left({}^3t_i - {}^2t_i\right)^3 \end{aligned} \tag{B.10}$$

$${}^4p_i - {}^3p_i = {}^3v_i \left({}^4t_i - {}^3t_i\right) - \frac{1}{6} J_i^{max} \left({}^4t_i - {}^3t_i\right)^3 \tag{B.11}$$

$$\begin{aligned} P_i^{trgt} - {}^4p_i = {} & {}^4v_i \left(t_i^{min} - {}^4t_i\right) + \frac{1}{2} a^{peak2} \left(t_i^{min} - {}^4t_i\right)^2 \\ & + \frac{1}{6} J_i^{max} \left(t_i^{min} - {}^4t_i\right)^3 \end{aligned} \tag{B.12}$$

As described in Chap. 4.2.1, we calculated one of the unknowns, a_i^{peak1}, by transforming this system of equations to a root-finding problem. By applying the modified Anderson-Björck-King method (cf. Appendix A), we attained the value for a_i^{peak1}. All following steps are trivial, but for reasons of completeness, they are presented. The values for the other eleven unknown variables are derived successively, and the trajectory parameters of $\mathcal{M}_i$, that is, ${}^l\vec{m}_i(t)$ with $l \in \{1, \ldots, 4\}$ and ${}^l\vartheta_i$ with $l \in \{1, \ldots, 4\}$, are calculated.

$$ {}^2t_i = T_i + \frac{a_i^{peak1} - A_i}{J_i^{max}} \tag{B.13} $$

$$ {}^3t_i = {}^2t_i + \frac{a_i^{peak1}}{J_i^{max}} \tag{B.14} $$

$$ {}^2v_i = \left(\frac{1}{2}\,A_i - a_i^{peak1}\right)\left({}^2t_i - T_i\right) + 2V_i \tag{B.15} $$

$$ {}^3v_i = \frac{1}{2}\,a_i^{peak1}\left({}^3t_i - {}^2t_i\right) + {}^2v_i \tag{B.16} $$

$$ \begin{aligned} {}^2p_i = {} & V_i\left({}^2t_i - T_i\right) + \frac{1}{2}\,A_i\left({}^2t_i - T_i\right)^2 \\ & + \frac{1}{6}\,J_i^{max}\left({}^2t_i - T_i\right)^3 + P_i \end{aligned} \tag{B.17} $$

$$ \begin{aligned} {}^3p_i = {} & {}^2p_i + {}^2v_i\left({}^3t_i - {}^2t_i\right) + \frac{1}{2}\,a_i^{peak1}\left({}^3t_i - {}^2t_i\right)^2 \\ & - \frac{1}{6}\,J_i^{max}\left({}^3t_i - {}^2t_i\right)^3 \end{aligned} \tag{B.18} $$

$$ a_i^{peak2} = -\sqrt{J_i^{max}\left(V_i^{trgt} - {}^3v_i\right)} \tag{B.19} $$

$$ {}^4t_i = {}^3t_i - \frac{a_i^{peak2}}{J_i^{max}} \tag{B.20} $$

$$ t_i^{min} = {}^4t_i - \frac{a_i^{peak2}}{J_i^{max}} \tag{B.21} $$

$$ {}^4v_i = \frac{1}{2}\,a_i^{peak2}\left({}^4t_i - {}^3t_i\right) + {}^3v_i \tag{B.22} $$

$$ {}^4p_i = {}^3p_i + {}^3v_i\left({}^4t_i - {}^3t_i\right) - \frac{1}{6}\,J_i^{max}\left({}^4t_i - {}^3t_i\right)^3 \tag{B.23} $$

Now the system of equations is completely and uniquely solved, such that we obtain the correct and desired time-optimal trajectory. As a final step, we have to parameterize the elements of $\mathcal{M}_i$.

$$ {}^1\vartheta_i = \left[T_i,\, {}^2t_i\right] \qquad \text{(cf. eqn. (3.9), p. 34)} \tag{B.24} $$

$$ {}^2\vartheta_i = \left[{}^2t_i,\, {}^3t_i\right] \tag{B.25} $$

$$^3\vartheta_i = \left[{}^3t_i, {}^4t_i\right] \tag{B.26}$$
$$^4\vartheta_i = \left[{}^4t_i, t_i^{min}\right] \tag{B.27}$$

$$^1\mathcal{V}_i = \left\{{}^1\vartheta_i\right\} \tag{B.28}$$
$$^2\mathcal{V}_i = \left\{{}^2\vartheta_i\right\} \tag{B.29}$$
$$^3\mathcal{V}_i = \left\{{}^3\vartheta_i\right\} \tag{B.30}$$
$$^4\mathcal{V}_i = \left\{{}^4\vartheta_i\right\} \tag{B.31}$$

$$^1j_i(t) = J_i^{max} \tag{B.32}$$
$$^2j_i(t) = -J_i^{max} \tag{B.33}$$
$$^3j_i(t) = -J_i^{max} \tag{B.34}$$
$$^4j_i(t) = J_i^{max} \tag{B.35}$$

$$^1a_i(t) = A_i + J_i^{max}\,(t - T_i) \tag{B.36}$$
$$^2a_i(t) = a_i^{peak1} - J_i^{max}\,(t - {}^2t_i) \tag{B.37}$$
$$^3a_i(t) = -J_i^{max}\,(t - {}^3t_i) \tag{B.38}$$
$$^4a_i(t) = a_i^{peak2} + J_i^{max}\,(t - {}^4t_i) \tag{B.39}$$

$$^1v_i(t) = V_i + A_i(t - T_i) + \frac{1}{2}\,J_i^{max}\,(t - T_i)^2 \tag{B.40}$$
$$^2v_i(t) = {}^1v_i\left({}^2t_i\right) + a_i^{peak1}\left(t - {}^2t_i\right) - \frac{1}{2}\,J_i^{max}\left(t - {}^2t_i\right)^2 \tag{B.41}$$
$$^3v_i(t) = {}^2v_i\left({}^3t_i\right) - \frac{1}{2}\,J_i^{max}\left(t - {}^3t_i\right)^2 \tag{B.42}$$
$$^4v_i(t) = {}^3v_i\left({}^4t_i\right) + a_i^{peak2}\left(t - {}^4t_i\right) + \frac{1}{2}\,J_i^{max}\left(t - {}^4t_i\right)^2 \tag{B.43}$$

$$^1p_i(t) = P_i + V_i(t - T_i) + \frac{1}{2}\,A_i(t - T_i)^2 + \frac{1}{6}\,J_i^{max}\,(t - T_i)^3 \tag{B.44}$$
$$^2p_i(t) = {}^1p_i\left({}^2t_i\right) + {}^1v_i\left({}^2t_i\right)\left(t - {}^2t_i\right) + \frac{1}{2}\,a_i^{peak1}\left(t - {}^2t_i\right)^2 - \frac{1}{6}\,J_i^{max}\left(t - {}^2t_i\right)^3 \tag{B.45}$$
$$^3p_i(t) = {}^2p_i\left({}^3t_i\right) + {}^2v_i\left({}^3t_i\right)\left(t - {}^3t_i\right) - \frac{1}{6}\,J_i^{max}\left(t - {}^3t_i\right)^3 \tag{B.46}$$

$$^4p_i(t) = {}^3p_i\left({}^4t_i\right) + {}^3v_i\left({}^4t_i\right)\left(t - {}^4t_i\right) + \frac{1}{2}\, a_i^{peak2}\left(t - {}^4t_i\right)^2 + \frac{1}{6}\, J_i^{max}\left(t - {}^4t_i\right)^3 \tag{B.47}$$

With

$$^l\vec{m}_i(t) = \left({}^lp_i(t), {}^lv_i(t), {}^la_i(t), {}^lj_i(t)\right) \quad \forall\, l \in \{1, \ldots, 4\} \tag{B.48}$$

a one-dimensional Type IV, Variant A trajectory at time instant T_i $\mathcal{M}_i$ is completely described (cf. eqn. (3.10), p. 34):

$$\mathcal{M}_i(t) = \left\{ \left({}^1\vec{m}_i(t), {}^1\mathcal{V}_i\right), \left({}^2\vec{m}_i(t), {}^2\mathcal{V}_i\right), \left({}^3\vec{m}_i(t), {}^3\mathcal{V}_i\right), \left({}^4\vec{m}_i(t), {}^4\mathcal{V}_i\right) \right\}. \tag{B.49}$$

Appendix C
Details on the *PosTriZeroNegTri* Acceleration Profile (Step 2)

C.1 Position-Error Function

Written form of the position-error function for the Step 2 *PosTriZeroNegTri* acceleration profile:

$$
{}^{PosTriZeroNegTri}p_i^{err2}\left(a_i^{peak1}\right) = \frac{1}{48\ ({}_kJ_i^{max})^3}\Bigg(-8\ ({}_kA_i)^3\ ({}_kJ_i^{max})
$$

$$
+48\,{}_kA_i\left({}_ka_i^{peak1}\right)^2({}_kJ_i^{max}) + 30\ ({}_kA_i)^2\Bigg(-80\ (t_i^{sync})\ ({}_kJ_i^{max})^2 + \sqrt{18}\,({}_kJ_i^{max})
$$

$$
\left(\sqrt{\left(-({}_kA_i)^2 + 2\left(\left({}_ka_i^{peak1}\right)^2 + ({}_kJ_i^{max})\ ({}_kV_i - ({}_kV_i^{trgt}))\right)\right)}\right.
$$

$$
\left.+\sqrt{-2\ ({}_kA_i)^2 + 4\left(\left({}_ka_i^{peak1}\right)^2 + ({}_kJ_i^{max})\ ({}_kV_i - ({}_kV_i^{trgt}))\right)}\right)\Bigg) - 6\Bigg(8\ \left({}_ka_i^{peak1}\right)^3
$$

$$
({}_kJ_i^{max}) + ({}_kJ_i^{max})\Bigg(-8\,({}_kJ_i^{max})^2\,({}_kP_i - {}_kP_i^{trgt} + (t_i^{sync})\,{}_kV_i) + ({}_kJ_i^{max})
$$

$$
\left({}_kV_i - ({}_kV_i^{trgt})\right)\sqrt{-2\ ({}_kA_i)^2 + 4\left({}_ka_i^{peak1}\right)^2 + 4\,({}_kJ_i^{max})\ ({}_kV_i - ({}_kV_i^{trgt}))} + \sqrt{18}
$$

$$
({}_kV_i - ({}_kV_i^{trgt}))\ ({}_kJ_i^{max})\sqrt{\left(-({}_kA_i)^2 + 2\left({}_ka_i^{peak1}\right)^2 + 2\,({}_kJ_i^{max})\ ({}_kV_i - ({}_kV_i^{trgt}))\right)}\Bigg)
$$

$$
+\left({}_ka_i^{peak1}\right)^2\Bigg(-8\ (t_i^{sync})\ ({}_kJ_i^{max})^2 + \sqrt{18}\,({}_kJ_i^{max})
$$

$$
\left(\sqrt{\left(-({}_kA_i)^2 + 2\left(\left({}_ka_i^{peak1}\right)^2 + ({}_kJ_i^{max})\ ({}_kV_i - ({}_kV_i^{trgt}))\right)\right)}\right.
$$

$$
\left.+\sqrt{-2\ ({}_kA_i)^2 + 4\left(\left({}_ka_i^{peak1}\right)^2 + ({}_kJ_i^{max})\ ({}_kV_i - ({}_kV_i^{trgt}))\right)}\right)\Bigg)\Bigg)\Bigg).
$$

C.2 Setting up the Parameters of $\mathcal{M}_i$

In order to achieve a completely described derivation of the Type IV on-line trajectory generation algorithm, the calculations of the unknown variables of eqns. (5.29) – (5.42) (Chap. 5.3.1, p. 88) are presented here. Afterwards, all parameters of the final trajectory $\mathcal{M}_i$ at time instant T_i are calculated for one DOF k.

Analogous to App. B.3 we again repeat the respective system of equations represented by eqns. (5.29) – (5.42).

$$ {}_k^2t_i - T_i = \frac{\left({}_ka^{peak1} - {}_kA_i\right)}{{}_kJ_i^{max}} \tag{C.1} $$

$$ {}_k^3t_i - {}_k^2t_i = \frac{{}_ka^{peak1}}{{}_kJ_i^{max}} \tag{C.2} $$

$$ {}_k^5t_i - {}_k^4t_i = -\frac{{}_ka^{peak2}}{{}_kJ_i^{max}} \tag{C.3} $$

$$ t_i^{sync} - {}_k^5t_i = -\frac{{}_ka^{peak2}}{{}_kJ_i^{max}} \tag{C.4} $$

$$ {}_k^2v_i - V_i = \frac{1}{2}\left({}_k^2t_i - T_i\right)\left({}_kA_i + {}_ka^{peak1}\right) \tag{C.5} $$

$$ {}_k^3v_i - {}_k^2v_i = \frac{1}{2}\left({}_k^3t_i - {}_k^2t_i\right)\, {}_ka^{peak1} \tag{C.6} $$

$$ {}_k^4v_i - {}_k^3v_i = 0 \tag{C.7} $$

$$ {}_k^5v_i - {}_k^4v_i = \frac{1}{2}\left({}_k^5t_i - {}_k^4t_i\right)\, {}_ka^{peak2} \tag{C.8} $$

$$ {}_kV_i^{trgt} - {}_k^5v_i = \frac{1}{2}\left(t_i^{sync} - {}_k^5t_i\right)\, {}_ka^{peak2} \tag{C.9} $$

$$ \begin{aligned} {}_k^2p_i - {}_kP_i = {}& {}_kV_i\left({}_k^2t_i - T_i\right) + \frac{1}{2}\,{}_kA_i\left({}_k^2t_i - T_i\right)^2 \\ & + \frac{1}{6}\,{}_kJ_i^{max}\left({}_k^2t_i - T_i\right)^3 \end{aligned} \tag{C.10} $$

$$ \begin{aligned} {}_k^3p_i - {}_k^2p_i = {}& {}_k^2v_i\left({}_k^3t_i - {}_k^2t_i\right) + \frac{1}{2}\,{}_ka_i^{peak1}\left({}_k^3t_i - {}_k^2t_i\right)^2 \\ & - \frac{1}{6}\,{}_kJ_i^{max}\left({}_k^3t_i - {}_k^2t_i\right)^3 \end{aligned} \tag{C.11} $$

$$ {}_k^4p_i - {}_k^3p_i = {}_k^3v_i\left({}_k^4t_i - {}_k^3t_i\right) \tag{C.12} $$

$$ {}_k^5p_i - {}_k^4p_i = {}_k^4v_i\left({}_k^5t_i - {}_k^4t_i\right) - \frac{1}{6}\,{}_kJ_i^{max}\left({}_k^5t_i - {}_k^4t_i\right)^3 \tag{C.13} $$

$$ \begin{aligned} {}_kP_i^{trgt} - {}_k^5p_i = {}& {}_k^5v_i\left(t_i^{sync} - {}_k^5t_i\right) + \frac{1}{2}\,{}_ka_i^{peak2}\left(t_i^{sync} - {}_k^5t_i\right)^2 \\ & + \frac{1}{6}\,{}_kJ_i^{max}\left(t_i^{sync} - {}_k^5t_i\right)^3 \end{aligned} \tag{C.14} $$

These 14 equations contain 14 unknown variables: ${}_k^2t_i$, ${}_k^3t_i$, ${}_k^4t_i$, ${}_k^5t_i$, ${}_k^2v_i$, ${}_k^3v_i$, ${}_k^4v_i$, ${}_k^5v_i$, ${}_k^2p_i$, ${}_k^3p_i$, ${}_k^4p_i$, ${}_k^5p_i$, ${}_ka_i^{peak1}$, and ${}_ka_i^{peak2}$. The calculation of the first unknown, ${}_ka_i^{peak1}$, was already described in Chap. 5.3.1 (p. 89). As in App. B.3, the following calculation of the other 13 unknown variables can be successively derived in a straightforward way:

$$ {}_k^2t_i = T_i + \frac{{}_ka_i^{peak1} - {}_kA_i}{{}_kJ_i^{max}} \tag{C.15} $$

$$ {}_k^3t_i = {}_k^2t_i + \frac{{}_ka_i^{peak1}}{{}_kJ_i^{max}} \tag{C.16} $$

$$ {}_k^2v_i = \left(\frac{1}{2}\,{}_kA_i - {}_ka_i^{peak1}\right)\left({}_k^2t_i - T_i\right) + 2\,{}_kV_i \tag{C.17} $$

$$ {}_k^3v_i = \frac{1}{2}\,{}_ka_i^{peak1}\,\left({}_k^3t_i - {}_k^2t_i\right) + {}_k^2v_i \tag{C.18} $$

$$ \begin{aligned} {}_k^2p_i = {} & {}_kV_i\,\left({}_k^2t_i - T_i\right) + \frac{1}{2}\,{}_kA_i\,\left({}_k^2t_i - T_i\right)^2 \\ & + \frac{1}{6}\,{}_kJ_i^{max}\,\left({}_k^2t_i - T_i\right)^3 + {}_kP_i \end{aligned} \tag{C.19} $$

$$ \begin{aligned} {}_k^3p_i = {} & {}_k^2p_i + {}_k^2v_i\,\left({}_k^3t_i - {}_k^2t_i\right) + \frac{1}{2}\,{}_ka_i^{peak1}\,\left({}_k^3t_i - {}_k^2t_i\right)^2 \\ & - \frac{1}{6}\,{}_kJ_i^{max}\,\left({}_k^3t_i - {}_k^2t_i\right)^3 \end{aligned} \tag{C.20} $$

$$ {}_ka_i^{peak2} = -\sqrt{{}_kJ_i^{max}\left({}_kV_i^{trgt} - {}_k^3v_i\right)} \tag{C.21} $$

$$ {}_k^5t_i = t_i^{sync} + \frac{{}_ka_i^{peak2}}{{}_kJ_i^{max}} \tag{C.22} $$

$$ {}_k^5v_i = {}_kV_i^{trgt} - \frac{1}{2}\,{}_ka_i^{peak2}\,\left(t_i^{sync} - {}_k^5t_i\right) \tag{C.23} $$

$$ \begin{aligned} {}_k^5p_i = {} & {}_kP_i^{trgt} - {}_k^5v_i\,\left(t_i^{sync} - {}_k^5t_i\right) - \frac{1}{2}\,{}_ka_i^{peak2}\,\left(t_i^{sync} - {}_k^5t_i\right)^2 \\ & - \frac{1}{6}\,{}_kJ_i^{max}\,\left(t_i^{sync} - {}_k^5t_i\right)^3 \end{aligned} \tag{C.24} $$

$$ {}_k^4t_i = {}_k^5t_i + \frac{{}_ka_i^{peak2}}{{}_kJ_i^{max}} \tag{C.25} $$

$$ {}_k^4v_i = {}_k^3v_i \tag{C.26} $$

$$ {}_k^4p_i = {}_k^5p_i - {}_k^4v_i\,\left({}_k^5t_i - {}_k^4t_i\right) + \frac{1}{6}\,{}_kJ_i^{max}\,\left({}_k^5t_i - {}_k^4t_i\right)^3 . \tag{C.27} $$

Therewith, we know the complete and unique solution of eqns. (C.1) – (C.14), such that we can finally set the trajectory parameters of $\mathcal{M}_i$, which describe motion of DOF k.

$$ {}_k^1\vartheta_i = \left[T_i, {}_k^2t_i\right] \tag{C.28} $$
$$ {}_k^2\vartheta_i = \left[{}_k^2t_i, {}_k^3t_i\right] \tag{C.29} $$
$$ {}_k^3\vartheta_i = \left[{}_k^3t_i, {}_k^4t_i\right] \tag{C.30} $$
$$ {}_k^4\vartheta_i = \left[{}_k^4t_i, {}_k^5t_i\right] \tag{C.31} $$
$$ {}_k^5\vartheta_i = \left[{}_k^5t_i, t_i^{sync}\right] \tag{C.32} $$

$$ {}_k^1j_i(t) = {}_kJ_i^{max} \tag{C.33} $$
$$ {}_k^2j_i(t) = -{}_kJ_i^{max} \tag{C.34} $$
$$ {}_k^3j_i(t) = 0 \tag{C.35} $$
$$ {}_k^4j_i(t) = -{}_kJ_i^{max} \tag{C.36} $$
$$ {}_k^5j_i(t) = {}_kJ_i^{max} \tag{C.37} $$

$$ {}_k^1a_i(t) = {}_kA_i + {}_kJ_i^{max}\,(t - T_i) \tag{C.38} $$
$$ {}_k^2a_i(t) = {}_ka_i^{peak1} - {}_kJ_i^{max}\left(t - {}_k^2t_i\right) \tag{C.39} $$
$$ {}_k^3a_i(t) = 0 \tag{C.40} $$
$$ {}_k^4a_i(t) = -{}_kJ_i^{max}\left(t - {}_k^4t_i\right) \tag{C.41} $$
$$ {}_k^5a_i(t) = {}_ka_i^{peak2} + {}_kJ_i^{max}\left(t - {}_k^5t_i\right) \tag{C.42} $$

$$ {}_k^1v_i(t) = {}_kV_i + {}_kA_i\,(t - T_i) + \frac{1}{2}\,{}_kJ_i^{max}\,(t - T_i)^2 \tag{C.43} $$
$$ {}_k^2v_i(t) = {}_k^1v_i\left({}_k^2t_i\right) + {}_ka_i^{peak1}\left(t - {}_k^2t_i\right) - \frac{1}{2}\,{}_kJ_i^{max}\left(t - {}_k^2t_i\right)^2 \tag{C.44} $$
$$ {}_k^3v_i(t) = {}_k^3v_i \tag{C.45} $$
$$ {}_k^4v_i(t) = {}_k^3v_i\left({}_k^4t_i\right) - \frac{1}{2}\,{}_kJ_i^{max}\left(t - {}_k^4t_i\right)^2 \tag{C.46} $$
$$ {}_k^5v_i(t) = {}_k^4v_i\left({}_k^5t_i\right) + {}_ka_i^{peak2}\left(t - {}_k^5t_i\right) + \frac{1}{2}\,{}_kJ_i^{max}\left(t - {}_k^5t_i\right)^2 \tag{C.47} $$

$$ \begin{aligned} {}_k^1p_i(t) = {}_kP_i + {}_kV_i\,(t - T_i) + \frac{1}{2}\,{}_kA_i\,(t - T_i)^2 \\ + \frac{1}{6}\,{}_kJ_i^{max}\,(t - T_i)^3 \end{aligned} \tag{C.48} $$

$$ {}_k^2p_i(t) = {}_k^1p_i\left({}_k^2t_i\right) + {}_k^1v_i\left({}_k^2t_i\right)\left(t - {}_k^2t_i\right) + \frac{1}{2}\,{}_ka_i^{peak1}\left(t - {}_k^2t_i\right)^2 $$

$$- \frac{1}{6} \, {}_kJ_i^{max} \left(t \; - \; {}_k^2t_i \right)^3 \tag{C.49}$$

$${}_k^3p_i(t) = {}_k^2p_i \left({}_k^3t_i \right) \; + \; {}_k^3v_i \left({}_k^3t_i \right) \left(t \; - \; {}_k^3t_i \right) \tag{C.50}$$

$${}_k^4p_i(t) = {}_k^3p_i \left({}_k^4t_i \right) \; + \; {}_k^3v_i \left({}_k^4t_i \right) \left(t \; - \; {}_k^4t_i \right) \; - \; \frac{1}{6} \, {}_kJ_i^{max} \left(t \; - \; {}_k^4t_i \right)^3 \tag{C.51}$$

$$\begin{aligned} {}_k^5p_i(t) = {}_k^4p_i \left({}_k^5t_i \right) \; + \; {}_k^4v_i \left({}_k^5t_i \right) \left(t \; - \; {}_k^5t_i \right) \; + \; \frac{1}{2} \, {}_ka_i^{peak2} \left(t \; - \; {}_k^5t_i \right)^2 \\ + \; \frac{1}{6} \, {}_kJ_i^{max} \left(t \; - \; {}_k^5t_i \right)^3 \end{aligned} \tag{C.52}$$

With the results of eqns. (C.28) – (C.52), we can finally set up all trajectory parameters for one single DOF k at time instant T_i

$${}^l\vec{m}_i(t) = \left({}^lp_i(t), {}^lv_i(t), {}^la_i(t), {}^lj_i(t) \right) \quad \forall \; l \; \in \; \{1, \ldots, 5\} \tag{C.53}$$

$${}_k^l\vartheta_i \in {}^l\mathcal{V}_i \quad \forall \; l \; \in \; \{1, \ldots, 5\} \; . \tag{C.54}$$

Equations (C.53) and (C.54) represent the DOF k in $\mathcal{M}_i$ (cf. eqns. (3.9) and (3.10), p. 34).

Appendix D
Type IV On-Line Trajectory Generation in Very Simple Terms

The purpose of this appendix chapter is to describe the Type IV on-line trajectory generation algorithm in a less scientific but more metaphoric and descriptive way by means of a very simple example. The chapter does not contain new information, but only illustrates a part of this book in a plain and comprehensive way.

D.1 The Rocket Car Example

In the figurative sense, we can consider a rocket car as depicted in Fig. D.1 for the representation of a single DOF of a system to be controlled by the on-line trajectory generator. We assume an idealized car, *massless* and *without any friction*. We only can control the car by turning a lever as shown in Fig. D.1. If the lever is at its rightmost position, the right rocket runs at full power, which corresponds to an acceleration with $-A_i^{max}$, the maximum available acceleration value at time T_i. If the lever is positioned exactly centered, both rockets are turned off, and the car is not accelerated in any direction. Analogously, the car will accelerate with $A_i = A_i^{max}$ if the lever is in its leftmost position. A_i^{max} is the power of our rocket motor. At first sight, this sounds easy, but there is an additional restriction: The car possesses a *maximum lever-turning velocity* J_i^{max}, that is, we are not allowed to turn the

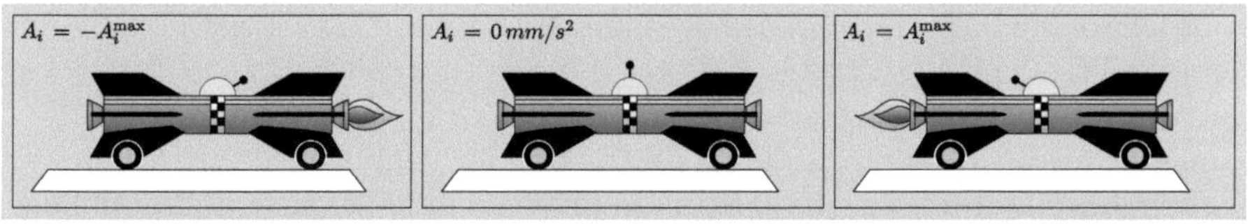

Fig. D.1 Illustration of the rocket car in three different acceleration states: $A_i = -A_i^{max}$, $A_i = 0\,mm/s^2$, and $A_i = A_i^{max}$.

lever jerkily. This leads to the consequence that the resulting acceleration progression is jump-free and the jerk of the car is limited.

D.2 Type IV OTG for One-DOF Systems

This appendix section takes up the approach for Type IV OTG from Chap. 4 (p. 45). Let us first transfer the problem formulation for the one-dimensional case to the rocket car example. Imagine that we work with time slices of T^{cycle} width.[1] Then, the time difference from one slice T_i to the next one T_{i+1} is T^{cycle}. At T_i, the algorithm gets the input values $\vec{W}_i$ as illustrated in Fig. D.2. To explain this more simply, it means that the car is at position P_i, drives with a velocity V_i, and accelerates with A_i. Summarized, this is the current state of motion $\vec{M}_i$. It is our task to reach the target position P_i^{trgt} in the shortest possible time. Furthermore, we want drive through P_i^{trgt} with a certain target velocity V_i^{trgt}. These two values, P_i^{trgt} and V_i^{trgt}, constitute the target state of motion $\vec{M}_i^{trgt}$. The target acceleration A_i^{trgt} has to be zero. It is our aim to transfer our car's current state of motion $\vec{M}_i$ to its target state of motion $\vec{M}_i^{trgt}$ in the shortest possible time t_i^{min}. Here, we have to take care for some motion constraints: Our car features a maximum velocity V_i^{max}, the power of its rocket motor is limited to A_i^{max}, and the maximum lever-turning velocity for controlling the rocket motor is J_i^{max}. These three boundary values are summarized in the vector $\vec{B}_i$.

Hence, the question here is: How can we calculate a trajectory that lets the car reach $\vec{M}_i^{trgt}$ in the shortest possible time? Because we want to be able to react to any unforeseen events instantaneously, for example, a thinking scientist could cross our way, we are only interested in the result for the time slice T_i. That means, even if we have to calculate the whole trajectory for reaching $\vec{M}_i^{trgt}$, we are only interested in the state of motion at the time slice T_{i+1}, that is, $\vec{M}_{i+1}$. As the last value, we have to consider S_i. S_i may be compared to a power on/off button for the OTG algorithm: If S_i is *one*, we have to calculate $\vec{M}_{i+1}$, and if S_i is *zero*, nobody is interested in our service, and we do not have to do anything. At the next time slice T_{i+1}, the algorithm starts over again and calculates the state of motion $\vec{M}_{i+2}$ without using any knowledge from the last time slice, because the input values could have unexpectedly changed.

In Chap. 3.2.2 (p. 39), we also distinguish between OTG Type IV, Variant *A* and OTG Type IV, Variant *B*. If we transfer this distinction to our rocket car, we can say that the Variant *A* algorithm works with constant boundary values $\vec{B}_i$ for all time slices. That means our maximum velocity, the power of the rocket motor, and the maximum lever-turning velocity do not change over time. The *B*-Variant is furthermore able to cope with changing elements of $\vec{B}_i$, that is, depending on the time slice, the power of the rocket motor

[1] A typical value for T^{cycle} is one millisecond.

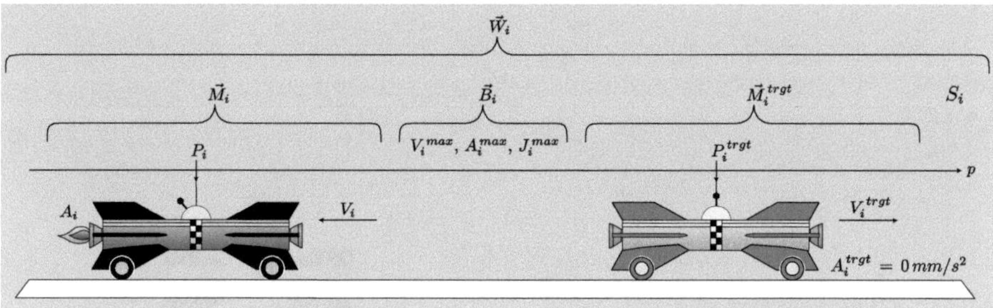

Fig. D.2 Illustration by means of the rocket car example: input values $\vec{W}_i$ at time T_i of the Type IV on-line trajectory generation algorithm for the one-dimensional case.

may change, the maximum velocity may change, or even the maximum lever-turning velocity may change.

In Chap. 4.2 (p. 49) the algorithm that solves this problem for Variant A and Variant B, is discussed with all details and nuances.

D.3 Type IV OTG for Multi-DOF Systems

We now extend the rocket car example of the previous section in order to apply the Type IV OTG algorithm to systems with multiple DOFs, for example, robots with multiple joints. We continue using the same non-scientific and plain way of presentation of the previous section.

Actually, we have to solve almost the same problem as in the one-dimensional case of Fig. D.2, but instead of only one rocket car, we now consider K independently driving rocket cars, as depicted in Fig. D.3. Each car is identified by the index k, and each car is assumed to be in a motion state ${}_k\vec{M}_i$, that is, the car with the index k is located at position ${}_kP_i$, at which it drives with a velocity ${}_kV_i$, and exerts a certain acceleration ${}_kA_i$. Analogous to the one-dimensional case, each car is assigned with a target state of motion ${}_k\vec{M}_i^{trgt}$ and some motion constraints ${}_k\vec{B}_i$. The aim of the Type IV on-line trajectory generation algorithm is to guide each car from its initial state of motion to its target state of motion in the shortest possible time — this is clear. But the additional challenge is to specify an algorithm that lets all K cars reach their target states of motion at the very same time: at the minimum possible synchronization time t_i^{sync}. Hence, the question to be answered here is: How can we calculate K trajectories, such that all K cars synchronously reach their state of motion in the shortest possible time?

As described in the previous section, each car can be turned on or off by the vector

$$\vec{S}_i = ({}_1S_i, \ldots, {}_kS_i, \ldots, {}_KS_i) \ , \qquad \text{(D.1)}$$

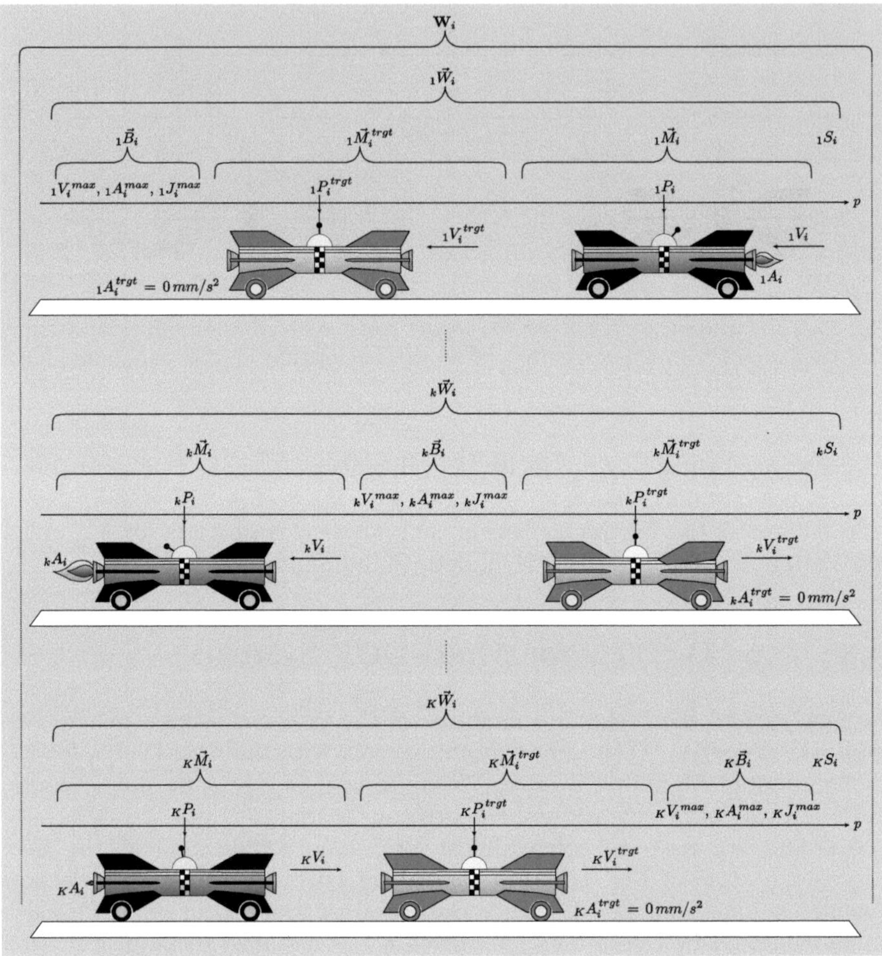

Fig. D.3 Illustration by means of the rocket car example: input values $\mathbf{W}_i$ of the Type IV on-line trajectory generation algorithm for a system with K independently acting DOFs at time T_i.

whose elements contain either a *one* or a *zero*. Hence, a rocket car is only taken into account if the corresponding element of the vector $\vec{S}_i$ equals *one*. If the respective value ${}_kS_i$ for the car with the index k is *zero*, the car will not be guided by the algorithm.

For a more compact embodiment of all these numbers, the matrix $\mathbf{W}_i$ is introduced at the top of Fig. D.3 and contains all necessary input variables of the OTG Type IV algorithm.

This algorithm is executed once per time slice, which again is represented by T_i; at each time slice three basic steps are executed:

1. Based on the input values $\mathbf{W}_i$, we calculate the minimum possible synchronization time t_i^{sync}. This time cannot be lower than the ride time of the rocket car with the greatest ride time.
2. We calculate a complete trajectory for each selected car, and we guarantee, that all selected cars will reach their target state of motion exactly at the time t_i^{sync}.
3. The output of the algorithm is calculated based on the result of Step 2 and only contains the state of motion for the current time slice.

Thus, we are able to react to unforeseen events (e.g., sudden slowdowns of cars in front of us) *immediately* after they are detected. The reaction time is not greater than T^{cycle}.

At the next time slice, T_{i+1}, we execute the same memoryless algorithm again. If nothing has happened, the previously calculated values ${}_1M_{i+1}, \ldots, {}_kM_{i+1}, \ldots, {}_KM_{i+1}$ are used as new input values for the time slice at T_{i+1}.

The difference between the A- and B-Variants is exactly the same as in the one-dimensional case, that is, Variant B is an extension of Variant A, which additionally can cope with varying values of ${}_1B_i, \ldots, {}_kB_i, \ldots, {}_KB_i$.

The algorithm, which can control the cars this way, is explained in detail in Chap. 5.3 (p. 80).

Remark D.1. *If written in the style of Chap. 5.3, the current motion states of all rocket cars can be simply summarized by*

$$\mathbf{M}_i = \left({}_1\vec{M}_i, \ldots, {}_k\vec{M}_i, \ldots, {}_K\vec{M}_i\right)^T , \tag{D.2}$$

and the same can easily be applied to the target states of motion and the motion constraints:

$$\mathbf{M}_i^{trgt} = \left({}_1\vec{M}_i^{trgt}, \ldots, {}_k\vec{M}_i^{trgt}, \ldots, {}_K\vec{M}_i^{trgt}\right)^T \tag{D.3}$$

$$\mathbf{B}_i = \left({}_1\vec{B}_i, \ldots, {}_k\vec{B}_i, \ldots, {}_K\vec{B}_i\right)^T . \tag{D.4}$$

Finally, we can express all input values $\mathbf{W}_i$ *by*

$$\begin{aligned}\mathbf{W}_i &= \left({}_1\vec{W}_i, \ldots, {}_k\vec{W}_i, \ldots, {}_K\vec{W}_i\right)^T \\ &= \left(\mathbf{M}_i, \mathbf{M}_i^{trgt}, \mathbf{B}_i, \vec{S}_i\right) .\end{aligned} \tag{D.5}$$

Abbreviations and Symbols

Abbreviations

BF	: Robot base frame
CCD	: Charge Coupled Device
CNC	: Computer numerical control
DOF	: Degree of freedom
EF	: External frame
HF	: Robot hand frame
HMI	: Human machine interface
MP	: Manipulation Primitive
NURBS	: Non-uniform, rational B-splines
OTG	: On-line trajectory generation
PRM	: Probabilistic road map
RRT	: Rapidly exploring random trees
RAMP	: Real-time adaptive motion planning
TF	: Task Frame
VGA	: Video graphics adapter

Symbols

Introduction (Chaps. 1 – 2)

$\vec{f}(t)$	: Forces and/or torques in joint space	(p. 2)
$\vec{p}(t)$	: Position in task space	(p. 3)
$\dot{\vec{p}}(t)$	: Velocity in task space	(p. 3)
$\ddot{\vec{p}}(t)$	: Acceleration in task space	(p. 3)
$\vec{p}_d(t)$	: Position set-point in task space	(p. 3)
$\dot{\vec{p}}_d(t)$	: Velocity set-point in task space	(p. 3)
$\ddot{\vec{p}}_d(t)$	: Acceleration set-point in task space	(p. 3)
$\vec{q}(t)$	: Position in joint space	(p. 2)

$\vec{\dot{q}}(t)$	: Velocity in joint space	(p. 2)
$\vec{\ddot{q}}(t)$	: Acceleration in joint space	(p. 2)
$\vec{q}_d(t)$	: Position set-point in joint space	(p. 2)
$\vec{\dot{q}}_d(t)$	: Velocity set-point in joint space	(p. 2)
$\vec{\ddot{q}}_d(t)$	: Acceleration set-point in joint space	(p. 2)
$\vec{s}(t)$	: Generic sensor signal in task space	(p. 3)
$\vec{s}_d(t)$	: Command variable in task space	(p. 3)

On-line trajectory generation (Chaps. 3 – 6 and 8 – 10)

$a_0, \ldots, a_4$	: Polynomial coefficients	(p. 34)
${}^l\vec{a}_i(t)$	: Vector of acceleration polynomials at T_i and segment l	(p. 34)
$\vec{A}_i$	: Acceleration vector at T_i	(p. 33)
${}_kA_i$	: Acceleration magnitude for DOF k vector at T_i	(p. 33)
$\vec{A}_i^{max}$	: Vector of maximum accelerations at T_i	(p. 34)
α	: Type-dependent integer value for the classification of OTG	(p. 38)
$\vec{B}_i$	: Motion constraints for one single DOF at T_i	(p. 45)
$\mathbf{B}_i$	: Matrix of boundary conditions for motion properties atT_i	(p. 34)
β	: Type-dependent integer value for the classification of OTG	(p. 38)
$\mathbb{B}$	: Set of binary numbers	(p. 37)
γ	: Vector with $\alpha - \beta - 2$ elements to check for colinearity	(p. 101)
${}_kD_i$	: Magnitude of the derivative of jerk for DOF k vector at T_i	(p. 33)
$\vec{D}_i$	: Vector of the derivative of jerk at T_i	(p. 33)
$\vec{D}_i^{max}$	: Vector of maximum derivatives of jerk at T_i	(p. 34)
δ	: Vector with β elements to check for colinearity	(p. 103)
${}^r\mathcal{D}_{Step1}$	: Domain for the system of equations of the motion profile ${}^r\Psi^{Step1}$	(p. 48)
h	: Index of a concrete intermediate motion state	(p. 161)
H	: Number of intermediate motion states	(p. 161)
$\mathcal{H}$	: Set of holes in the $(\alpha + 1)$-dimensional space	(p. 76)
i	: Index for one time instant T_i	(p. 33)
${}_k\mathcal{I}_i$	: Set of time instants for a DOF k at T_i	(p. 73)
${}^l\vec{j}_i(t)$	: Vector of jerk polynomials at T_i and segment l	(p. 34)
$\vec{J}_i$	: Jerk vector at T_i	(p. 33)
$\vec{J}_i^{max}$	: Vector of maximum jerks at T_i	(p. 34)
${}_kJ_i$	: Jerk magnitude for DOF k vector at T_i	(p. 33)
k	: One single DOFs with $k \in \{1, \ldots, K\}$	(p. 33)
K	: Number of DOFs	(p. 33)
κ	: One single DOFs with the index $\kappa \in \{1, \ldots, K\}$	(p. 99)
l	: One single trajectory segment with $l \in \{1, \ldots, L\}$	(p. 34)

L	: Maximum number of trajectory segments	(p. 34)
λ	: Number of control cycles for parameter adaptation	(p. 165)
Λ	: Number of intermediate trajectory segments	(p. 46)
${}_k^l\vec{m}_i(t)$	: Vector of motion polynomials of segment l and DOF k at time T_i	(p. 34)
${}^l\mathbf{m}_i(t)$	: Matrix of motion polynomials of segment l at time T_i	(p. 34)
$\vec{M}_i$	: State of motion for one single DOF at T_i	(p. 45)
$\mathbf{M}_i$	: Motion state matrix at T_i	(p. 33)
$\mathbf{M}_i^{trgt}$	: Matrix containing the target state of motion a T_i	(p. 34)
$\mathcal{M}_i(t)$	: Parameterized trajectory at time T_i	(p. 34)
$\mathbb{N}$	: Set of natural numbers	(p. 49)
ν	: Number of cycles in-between two (visual servo control) set-points	(p. 159)
${}^l\vec{p}_i(t)$	: Vector of position polynomials at T_i and segment l	(p. 34)
${}_kP_i$	: Position magnitude for DOF k at T_i	(p. 33)
$\vec{P}_i$	: Position vector at T_i	(p. 33)
${}^r\Psi^{Step1}$	: Motion profile r for Step 1 (element of $\mathcal{P}_{Step1}$)	(p. 47)
$\mathcal{P}_{Step1}$	: Set of motion profiles for Step 1	(p. 47)
r	: Index for Step 1 motion profiles	(p. 47)
R	: Total number of elements of $\mathcal{P}_{Step1}$	(p. 47)
$\vec{\rho}_i$	: Ratio vector for homothety at T_i	(p. 99)
$\mathbb{R}$	: Set of real numbers	(p. 39)
s	: Index for Step 2 motion profiles	(p. 76)
S	: Total number of elements of $\mathcal{P}_{Step2}$	(p. 76)
$\vec{S}_i$	: Selection vector at T_i	(p. 37)
$\mathcal{S}$	: Intersection of all Step 2 input domains	(p. 78)
Δt	: Value for polynomial time-shifting	(p. 34)
${}_kt_i^{min}$	: Minimum possible execution time for DOF k calculated at T_i	(p. 46)
t_i^{sync}	: Synchronization time calculated at T_i	(p. 35)
T^{cycle}	: Control cycle time	(p. 33)
$\mathcal{T}$	: Set of time instants	(p. 33)
T_0, T_i, T_N	: Discrete time instants of $\mathcal{T}$	(p. 33)
${}_k\tilde{v}_i$	: Velocity achieved by a peak-peak accel. profile for DOF k at T_i	(p. 91)
${}^l\vec{v}_i(t)$	: Vector of velocity polynomials at T_i and segment l	(p. 34)
${}_kV_i$	: Velocity magnitude for DOF k vector at T_i	(p. 33)
$\vec{V}_i$	: Velocity vector at T_i	(p. 33)
$\vec{V}_i^{max}$	: Vector of maximum velocities at T_i	(p. 34)
${}_k^l\vartheta_i$	: Time interval for segment l and DOF k at time T_i	(p. 34)
${}^l\mathcal{V}_i$	: Set of time intervals for segment l at time T_i	(p. 34)
$\vec{W}_i$	: OTG input values for one single DOF at T_i	(p. 45)
$\mathbf{W}_i$	: All input parameters of the OTG algorithm	(p. 37)
${}_k^z\zeta_i$	: z-th inoperative time interval for DOF k at T_i	(p. 71)

${}_k\mathcal{Z}_i$: Set of inoperative time intervals for DOF k at T_i (p. 70)
$\mathbb{Z}$: Set of integer numbers (p. 39)

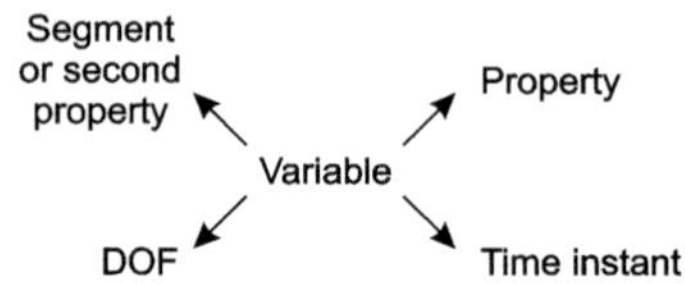

Fig. S.1 Convention for sub- and superscripts of all variables except sets and profiles.

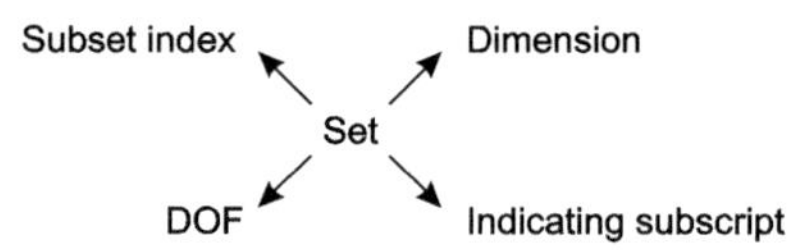

Fig. S.2 Convention for sub- and superscripts of sets.

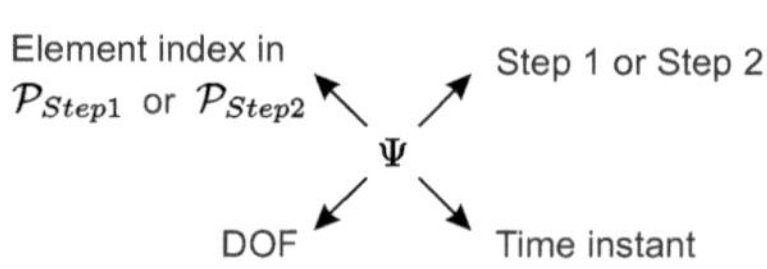

Fig. S.3 Convention for sub- and superscripts of profiles.

Hybrid Switched-System Control (Chap. 7)

ANC_i : Anchor frame of the Task Frame at T_i (p. 110)
BF : Robot base frame (updated by the system every control cycle) (p. 110)
$\mathbb{B}$: Set of binary numbers (p. 119)
c : Index for one control submodule $c \in \mathcal{C}$ (p. 110)
$\mathcal{C}$: Set of all m control submodules in the hybrid switched-system (p. 111)
${}^l_kD^c_i$: One single set-point for controller c, level l, and DOF k at T_i (p. 110)
$\mathcal{D}_i$: Set-point set of a hybrid move command $\mathcal{HM}_i$ at T_i (p. 110)
EF : External frame (updated by the system every control cycle) (p. 110)
$\mathbf{E}_i$: Flag assignment matrix at T_i (p. 122)
$\vec{f}^c_i$: Availability flag vector for controller c at T_i (p. 120)
FFC_i : Frame for feedforward control of the Task Frame at T_i (p. 110)
$\mathbf{F}^c_i$: Availability matrix for controller c at T_i (p. 120)

$\mathbf{G}_i^r$	: Control variable assignment matrix with $r \in \{pos, vel, acc\}$ at T_i	(p. 121)
$\mathbf{H}_i$	: Adaptive selection matrix at time T_i $\mathcal{MP}_i$ at T_i	(p. 121)
HF	: Hand frame (updated by the system every control cycle)	(p. 110)
$\mathcal{HM}_i$	: Parameter set of a hybrid move command of $\mathcal{MP}_i$ at T_i	(p. 109)
i	: Index for one time instant T_i	(p. 109)
I	: Identity matrix	(p. 117)
k	: Index for one DOF $k \in \mathcal{K}$	(p. 110)
$\mathcal{K}$	: Set of all DOFs $\{x, y, z, \textcircled{x}, \textcircled{y}, \textcircled{z}\}$	(p. 111)
l	: Index for one level $l \in \mathcal{L}$	(p. 110)
λ_i	: Stop condition of a manipulation primitive $\mathcal{MP}_i$ at T_i	(p. 109)
$\mathcal{L}$	: Set of all control m levels	(p. 111)
m	: Number of control submodules in the hybrid switched-system	(p. 111)
${}^A\mathbf{M}_i^B$	: Motion state of frame B w.r.t. frame A at T_i	(p. 111)
$\mathcal{MP}_i$	: Set of Manipulation Primitive parameters at T_i	(p. 109)
n	: Number of sensor systems	(p. 112)
${}^r\vec{o}_i^{\,c}$	: Control variable vectors with $r \in \{pos, vel, acc\}$	(p. 120)
${}^r\mathbf{O}_i^c$	: Controller output matrices with $r \in \{pos, vel, acc\}$	(p. 120)
RF_i	: Reference frame of the Task Frame at T_i	(p. 110)
$\mathbb{R}$	: Set of real numbers	(p. 110)
$\mathbf{S}_i^c$	: Classical selection matrix for one controller c	(p. 117)
$\mathcal{S}$	: Set of sensor signals	(p. 123)
τ_i	: Tool command of a manipulation primitive $\mathcal{MP}_i$ at T_i	(p. 109)
$\vec{\theta}_i$	: Cartesian pose vector with six elements: ${}_x\theta_i, {}_y\theta_i, {}_z\theta_i, {}_{\textcircled{x}}\theta_i, {}_{\textcircled{y}}\theta_i, {}_{\textcircled{z}}\theta_i$	(p. 110)
$\mathcal{TF}_i$	: Task Frame parameters of a hybrid move command $\mathcal{HM}_i$ at T_i	(p. 110)
WF	: World frame (updated by the system every control cycle)	(p. 110)
$\mathbf{Z}_i^c$	: Allocation matrix for controller c at T_i	(p. 119)

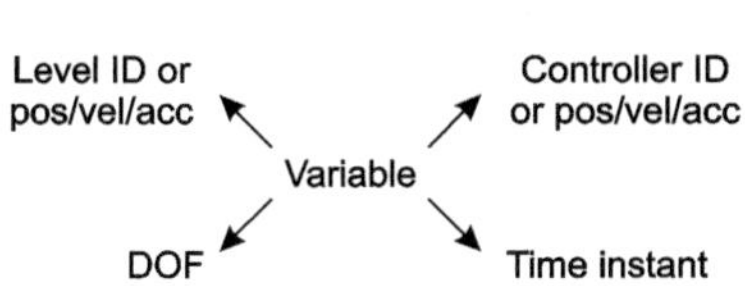

Fig. S.4 Convention for sub- and superscripts of variables and values in a hybrid switched-system.

References

[1] ABB AB Robotics Products, SE-721 68 Västerås, Sweden. *Product Specification, Controller IRC5 with FlexPendant, RobotWare RW 5.11*, 2008. 3HAC021785-001, Revision L.

[2] ABB AB Robotics Products, SE-721 68 Västerås, Sweden. *Product Specification, Controller Software IRC5, RobotWare 5.11*, 2008. 3HAC022349-001, Revision J.

[3] ABB AB Robotics Products, SE-721 68 Västerås, Sweden. Homepage. `http://www.abb.com/robotics` (accessed: Dec. 15, 2008). Internet, 2008.

[4] ABB Automation Technologies AB Robotics, SE-721 68 Västerås, Sweden. *Product Specification PickMaster*, 2008. 3HAC 5842-10, Version 3.20.

[5] K. Ahn, W. K. Chung, and Y. Yourn. Arbitrary states polynomial-like trajectory (ASPOT) generation. In *Proc. of the 30th Annual Conference of IEEE Industrial Electronics Society*, volume 1, pages 123–128, Busan, South Korea, November 2004.

[6] A. Albu-Schäffer, C.Ott, and G. Hirzinger. A unified passivity-based control framework for position, torque and impedance control of flexible joint robots. *The International Journal of Robotics Research*, 26(1):23–39, January 2007.

[7] A. Albu-Schäffer and G. Hirzinger. Cartesian impedance control techniques for torque controlled light-weight robots. In *Proc. of the IEEE International Conference on Robotics and Automation*, volume 1, pages 657–663, Washington, D.C., USA, May 2002.

[8] A. Albu-Schäffer, C. Ott, U. Frese, and G. Hirzinger. Cartesian impedance control of redundant robots: Recent results with the DLR-light-weight-arms. In *Proc. of the IEEE International Conference on Robotics and Automation*, volume 3, pages 3704–3709, Taipei, Taiwan, September 2003.

[9] T. Amemiya and T. Maeda. Asymmetric oscillation distorts the perceived heaviness of handheld objects. *IEEE Trans. on Haptics*, 1(1):9–18, January 2008.

[10] C. H. An, C. G. Atkeson, and J. M. Hollerbach. Experimental determination of the effect of feedforward control on trajectory tracking errors. In *Proc. of the IEEE International Conference on Robotics and Automation*, volume 3, pages 55–60, San Francisco, CA, USA, April 1986.

[11] C. H. An, C. G. Atkeson, and J. M. Hollerbach. *Model-Based Control of Robot Manipulator.* MIT Press, 1988.

[12] N. Anderson and Å. Björck. A new high order method of regula falsi type for computing a root of an equation. *BIT Numerical Mathematics*, 13(3):253–264, September 1973.

[13] R. L. Andersson. *A Robot Ping-Pong Player: Experiment in Real-Time Intelligent Control.* MIT Press, Cambridge, MA, USA, March 1988.

[14] R. L. Andersson. Aggressive trajectory generator for a robot ping-pong player. *IEEE Control Systems Magazine*, 9(2):15–21, February 1989.

[15] ATI Industrial Automation, Inc., 1031 Goodworth Dr. Apex, NC, 27539, USA. Homepage. `http://www.ati-ia.com` (accessed: Dec. 15, 2008). Internet, 2008.

[16] J. Baeten and J. De Schutter. *Integrated Visual Servoing and Force Control*, volume 8 of *Springer Tracts in Advanced Robotics*. Springer, 2004.

[17] J. Barbic and D. L. James. Six-dof haptic rendering of contact between geometrically complex reduced deformable models. *IEEE Trans. on Haptics*, 1(1):39–52, January 2008.

[18] Battenberg ROBOTIC GmbH & Co.KG, Zum Stempel 11, D-35043 Marburg, Germany. Homepage. `http://www.battenberg.biz` (accessed: Dec. 15, 2008). Internet, 2008.

[19] B. Bäuml and G. Hirzinger. Agile robot development (aRD): A pragmatic approach to robotic software. In *Proc. of the IEEE/RSJ International Conference on Intelligent Robots and Systems*, pages 3741–3748, Beijing, China, October 2006.

[20] S. A. Bazaz and B. Tondu. On-line computing of a robotic manipulator joint trajectory with velocity and acceleration constraints. In *Proc. of the IEEE International Symposium on Assembly and Task Planning*, pages 1–6, Marina del Rey, CA, USA, August 1997.

[21] S. A. Bazaz and B. Tondu. Minimum time on-line joint trajectory generator based on low order spline method for industrial manipulators. *Robotics and Autonomous Systems*, 29(4):3–17, December 1999.

[22] Bernecker + Rainer Industrie Elektronik Ges.m.b.H., B&R Straße 1, D-5142 Eggelsberg, Germany. Homepage. `http://www.br-automation.com` (accessed: Dec. 15, 2008). Internet, 2008.

[23] Y. Bestaoui. On-line motion generation with velocity and acceleration constraints. *Robotics and Autonomous Systems*, 5(3):279–288, November 1989.

[24] L. Biagiotti and C. Melchiorri. *Trajectory Planning for Automatic Machines and Robots.* Springer, Berlin, Heidelberg, Germany, first edition, 2008.

[25] L. Biagiotti and C. Melchiorri. *Trajectory Planning for Automatic Machines and Robots*, chapter 3, Composition of Elementary Trajectories, pages 59–150. Springer, Berlin, Heidelberg, Germany, first edition, 2008.

[26] C. G. L. Bianco and A. Piazzi. Minimum-time trajectory planning of mechanical manipulators under dynamic constraints. *International Journal of Control*, 75(13):967–980, 2002.

[27] J. E. Bobrow. Optimal robot path planning using the minimum-time criterion. *IEEE Trans. on Robotics and Automation*, 4(4):443–450, August 1988.

[28] J. E. Bobrow, S. Dubowsky, and J. S. Gibson. Time-optimal control of robotic manipulators along specified paths. *The International Journal of Robotics Research*, 4(3):3–17, Fall 1985.

[29] M. Brady. Trajectory planning. In M. Brady, J. M. Hollerbach, T. L. Johnson, T. Lozano-Pérez, and M. T. Mason, editors, *Robot Motion: Planning and Control*, chapter 4, pages 221–243. MIT Press, 1982.

[30] M. S. Branicky. *Studies in Hybrid Systems: Modeling, Analysis, and Control.* PhD thesis, Electrical Engineering and Computer Science Dept., Massachusetts Institute of Technology, `http://dora.cwru.edu/msb/pubs.html` (accessed: Dec. 15, 2008), 1995.

[31] M. S. Branicky. Multiple Lyapunov functions and other analysis tools for switched and hybrid systems. *IEEE Trans. on Automatic Control*, 43(4):475–482, April 1998.

[32] M. S. Branicky. Stability of hybrid systems: State of the art. In *Proc. of the IEEE Conference on Decision and Control*, volume 1, pages 120–125, San Diego, CA, USA, December 1999.

[33] O. Brock. *Generation of Robot Motion: The Integration of Planning and Execution.* PhD thesis, Department of Computer Science, Stanford University, 1999.

[34] O. Brock and L. E. Kavraki. Decomposition-based motion planning: A framework for real-time motion planning in high-dimensional configuration spaces. In *Proc. of the IEEE International Conference on Robotics and Automation*, pages 1469–1474, Seoul, South Korea, May 2001.

[35] O. Brock and O. Khatib. Elastic strips: A framework for motion generation in human environments. *The International Journal of Robotics Research*, 21(12):1031–1052, December 2002.

[36] O. Brock, J. Kuffner, and J. Xiao. Manipulation for robot tasks. In B. Siciliano and O. Khatib, editors, *Springer Handbook of Robotics*, chapter 26, pages 615–645. Springer, Berlin, Heidelberg, Germany, first edition, 2008.

[37] B. Brogliato. *Nonsmooth Mechanics.* Communications and Control Engineering. Springer, London, UK, 1999.

[38] X. Broquère, D. Sidobre, and I. Herrera-Aguilar. Soft motion trajectory planner for service manipulator robot. In *Proc. of the IEEE/RSJ International Conference on Intelligent Robots and Systems*, pages 2808–2813, Nice, France, September 2008.

[39] H. Bruyninckx. *Kinematic Models for Robot Compliant Motion with Identification of Uncertainities.* PhD thesis, KU Leuven, Department of Mechanical Engineering, 1995.

[40] H. Bruyninckx and J. De Schutter. Specification of force-controlled actions in the task frame formalism — A synthesis. *IEEE Trans. on Robotics and Automation*, 12(4):581–589, August 1996.

[41] B. Cao, G. I. Dodds, and G. W. Irwin. Time-optimal and smooth constrained path planning for robot manipulators. In *Proc. of the IEEE International Conference on Robotics and Automation*, volume 3, pages 1853–1858, San Diego, CA, USA, May 1994.

[42] B. Cao, G. I. Dodds, and G. W. Irwin. A practical approach to near time-optimal inspection-task-sequence planning for two cooperative industrial robot arms. *The International Journal of Robotics Research*, 17(8):858–867, August 1998.

[43] R. H. Castain and R. P. Paul. An on-line dynamic trajectory generator. *The International Journal of Robotics Research*, 3(1):68–72, March 1984.

[44] S. Chand and K. L. Doty. On-line polynomial trajectories for robot manipulators. *The International Journal of Robotics Research*, 4(2):38–48, Summer 1985.

[45] F. Chaumentte and S. A. Hutchinson. Visual servo control. Part I: Basic approaches. *IEEE Robotics and Automation Magazine*, 4(13):82–90, December 2006.

[46] F. Chaumentte and S. A. Hutchinson. Visual servo control. Part II: Advanced approaches. *IEEE Robotics and Automation Magazine*, 1(14):109–118, March 2007.

[47] F. Chaumentte and S. A. Hutchinson. Visual servoing and visual tracking. In B. Siciliano and O. Khatib, editors, *Springer Handbook of Robotics*, chapter 24, pages 563–583. Springer, Berlin, Heidelberg, Germany, first edition, 2008.

[48] C.-Y. Chen, P.-S. Liao, C.-C. Cheng, and G.-F. Jong. Design and implementation of real-time nurbs interpolator for motion control. In *Proc. of the second IEEE Conference Industrial Electronics and Applications*, pages 426–431, Harbin, China, May 2007.

[49] Y. Chen and A. A. Desrochers. Structure of minimum-time control law for robotic manipulators with constrained paths. In *Proc. of the IEEE International Conference on Robotics and Automation*, volume 2, pages 971–976, Scottsdale, AZ, USA, May 1989.

[50] Y. Chen and A. A. Desrochers. A proof of the structure of the minimum-time control law of robotic manipulators using a hamiltonian formulation. *IEEE Trans. on Robotics and Automation*, 6(3):388–393, June 1990.

[51] Y. Chen and A. A. Desrochers. Minimum-time control laws for robotic manipulators. *International Journal of Control*, 57(1):1–27, January 1993.

[52] M.-Y. Cheng, M.-C. Tsai, and J.-C. Kuo. Real-time NURBS command generators for CNC servo controllers. *International Journal of Machine Tools and Manufacture*, 42(7):801–813, May 2002.

[53] S. Chiaverini and L. Sciavicco. The parallel approach to force/position control of robotic manipulators. *IEEE Trans. on Robotics and Automation*, 9(4):361–373, August 1993.

[54] H.-Y. Chuang and K.-H. Chien. A real-time NURBS motion interpolator for position control of a slide equilateral triangle parallel manipulator. *The International Journal of Advanced Manufacturing Technology*, 34(7):724–735, October 2007.

[55] W. Chung, L.-C. Fu, and S.-H. Hsu. Motion control. In B. Siciliano and O. Khatib, editors, *Springer Handbook of Robotics*, chapter 6, pages 133–159. Springer, Berlin, Heidelberg, Germany, first edition, 2008.

[56] D. Chwa, J. Kang, and J. Y. Choi. Online trajectory planning of robot arms for interception of fast maneuvering object under torque and velocity constraints. *IEEE Trans. on Systems, Man, and Cybernetics, Part A: Systems and Humans*, 35(6):831–843, November 2005.

[57] COMAU S.p.A. Robotics, Via Rivalta, 30, 10095, Grugliasco (Turin), Italy. *C4G Instruction Handbook Motion Programming, System Software Rel. 3.1x*, 2008. CR00757507_en-01/0208.

[58] COMAU S.p.A. Robotics, Via Rivalta, 30, 10095, Grugliasco (Turin), Italy. *C4G OPEN Instruction Handbook, System Software Rel. 3.1x*, 2008. CR00757550_en-03/0908.

[59] COMAU S.p.A. Robotics, Via Rivalta, 30, 10095, Grugliasco (Turin), Italy. *C4G Open, the Industrial Robots Open Control System for Universities and SMEs (Product Brochure)*, 2008.
[60] COMAU S.p.A. Robotics, Via Rivalta, 30, 10095, Grugliasco (Turin), Italy. Homepage. `http://www.comau.com` (accessed: Dec. 15, 2008). Internet, 2008.
[61] D. Constantinescu and E. A. Croft. Smooth and time-optimal trajectory planning for industrial manipulators along specified paths. *Journal of Robotic Systems*, 17(5):233–249, May 2000.
[62] J. J. Craig. *Introduction to Robotics: Mechanics and Control.* Prentice Hall, third edition, 2003.
[63] O. Dahl. Path-constrained robot control with limited torques—Experimental evaluation. *IEEE Trans. on Robotics and Automation*, 10(5):658–669, October 1994.
[64] O. Dahl and L. Nielsen. Torque limited path following by on-line trajectory time scaling. In *Proc. of the IEEE International Conference on Robotics and Automation*, volume 2, pages 1122–1128, Scottsdale, AZ, USA, May 1989.
[65] W. C. Davidon. Variable metric method for minimization. Argonne National Laboratory Research and Development Report 5990, May 1959. Republished in: SIAM Journal of Optimization, 1(1):1–17, Febuary 1991.
[66] W. P. Dayawansa and C. F. Martin. A converse Lyapunov theorem for a class of dynamical systems which undergo switching. *IEEE Trans. on Automatic Control*, 44(4):751–760, April 1999.
[67] R. A. DeCarlo, M. S. Branicky, S. Pettersson, and B. Lennartson. Perspectives and results on the stability and stabilizability of hybrid systems. *Proc. of the IEEE*, 88(7):1069–1082, July 2000.
[68] R. Diestel. *Graph Theory*, volume 173 of *Graduate Texts in Mathematics.* Springer, Heidelberg, Germany, thrid edition, 2005.
[69] J. Dong, P. M. Ferreira, and J. A. Stori. Feed-rate optimization with jerk constraints for generating minimum-time trajectories. *International Journal of Machine Tools and Manufacture*, 47(12–13):1941–1955, 2007.
[70] M. Dowell and P. Jarratt. A modified regula falsi method for computing the root of an equation. *BIT Numerical Mathematics*, 11(2):168–174, June 1971.
[71] M. Dowell and P. Jarratt. The "Pegasus" method for computing the root of an equation. *BIT Numerical Mathematics*, 12(4):503–508, December 1972.
[72] J. Duffy. The fallacy of modern hybrid control theory that is based on "orthogonal complements" of twist and wrench spaces. *Journal of Robotic Systems*, 7(2):139–144, April 1990.
[73] G. Engeln-Müllges and F. Uhlig. *Numerical Algorithms with C.* Springer, 1996.
[74] EtherCAT Technology Group, Ostendstraße 196, D-90482 Nuremberg, Germany. Homepage. `http://www.ethercat.org` (accessed: Dec. 15, 2008). Internet, 2008.
[75] European Robotics Research Network (EURON). Hompepage. `http://www.euron.org` (accessed: Dec. 15, 2008). Internet, 2008.
[76] FANUC Robotics Deutschland GmbH, Bernhäuser Straße 36, D-73765 Neuhausen a. d. F., Germany. *FANUC Force Sensor FS-10iA data sheet*, 2008.
[77] FANUC Robotics Deutschland GmbH, Bernhäuser Straße 36, D-73765 Neuhausen a. d. F., Germany. *FANUC iRVISION data sheet*, 2008.

[78] FANUC Robotics Deutschland GmbH, Bernhäuser Straße 36, D-73765 Neuhausen a. d. F., Germany. *FANUC Roboterserie, R-30iA Mate Steuerung, Bedienungshandbuch*, 2008. B-82724GE-1/01.

[79] FANUC Robotics Deutschland GmbH, Bernhäuser Straße 36, D-73765 Neuhausen a. d. F., Germany. Homepage. `http://www.fanuc.co.jp/en` (accessed: Dec. 15, 2008). Internet, 2008.

[80] G. Farmelo, editor. *It Must Be Beautiful: Great Equations of Modern Science*. Granta Publications, London, UK, 2002.

[81] R. Featherstone. The calculation of robot dynamics using articulated-body inertias. *The International Journal of Robotics Research*, 2(1):13–30, 1983.

[82] R. Featherstone. *Rigid Body Dynamics Algorithms*. Springer, New York, NY, USA, first edition, 2007.

[83] R. Featherstone and D. E. Orin. Dynamics. In B. Siciliano and O. Khatib, editors, *Springer Handbook of Robotics*, chapter 2, pages 35–65. Springer, Berlin, Heidelberg, Germany, first edition, 2008.

[84] J. T. Feddema and O. R. Mitchell. Vision-guided servoing with feature-based trajectory generation. *IEEE Trans. on Robotics and Automation*, 5(5):691–700, 1989.

[85] B. Finkemeyer. *Robotersteuerungsarchitektur auf der Basis von Aktionsprimitiven (in German)*. Shaker Verlag, Aachen, Germany, 2004.

[86] B. Finkemeyer, T. Kröger, D. Kubus, M. Olschewski, and F. M. Wahl. MiRPA: Middleware for robotic and process control applications. In *Workshop on Measures and Procedures for the Evaluation of Robot Architectures and Middleware at the IEEE/RSJ International Conference on Intellegent Robots and Systems*, pages 78–93, San Diego, CA, USA, October 2007.

[87] B. Finkemeyer, T. Kröger, and F. M. Wahl. Placing of objects in unknown environments. In *Proc. of the IEEE International Conference on Methods and Models in Automation and Robotics*, pages 975–980, Międzyzdroje, Poland, August 2003.

[88] B. Finkemeyer, T. Kröger, and F. M. Wahl. Executing assembly tasks specified by manipulation primitive nets. *Advanced Robotics*, 19(5):591–611, June 2005.

[89] P. Fiorini and Z. Shiller. Time optimal trajectory planning in dynamic environments. In *Proc. of the IEEE International Conference on Robotics and Automation*, volume 2, pages 1553–1558, Minneapolis, MN, USA, April 1996.

[90] T. Flash and E. Henis. Arm trajectory modifications during reaching towards visual targets. *Journal of Cognitive Neuroscience*, 3(3):220–230, Summer 1991.

[91] T. Flash and N. Hogan. The coordination of arm movements: An experimentally confirmed mathematical model. *Journal of Neuroscience*, 5:1688–1703, Summer 1985.

[92] R. Fletcher and M. J. D. Powell. A rapidly convergent descent method for minimization. *The Computer Journal*, 6(2):163–168, Summer 1963.

[93] D. A. Forsyth and J. Ponce. *Computer Vision A Modern Approach*. Pearson Education, 2003.

[94] K. S. Fu, R. C. Gonzalez, and C. S. G. Lee. *Robotics: Control, Sensing, Vision and Intelligence*. McGraw-Hill, Singapore, second edition, 1988.

[95] N. R. Gans. *Hybrid Switched System Visual Servo Control.* PhD thesis, Department of General Engineering, University of Illinois at Urbana-Champaign, 2005.
[96] N. R. Gans and S. A. Hutchinson. A switching approach to visual servo control. In *Proc. of the IEEE International Symposium on Intelligent Control*, pages 770–776, Vancouver, Canada, October 2002.
[97] N. R. Gans and S. A. Hutchinson. Stable visual servoing through hybrid switched-system control. *IEEE Trans. on Robotics*, 23(3):530–540, June 2007.
[98] T. Gat-Falik and T. Flash. A technique for time-jerk optimal planning of robot trajectories. *IEEE Trans. on Systems, Man, and Cybernetics, Part B: Cybernetics*, 29(1):83–95, February 1999.
[99] H. Geering, L. Guzzella, S. Hepner, and C. Onder. Time-optimal motions of robots in assembly tasks. *IEEE Trans. on Automatic Control*, 31(6):512–518, June 1986.
[100] GNU Scientific Library. Hompepage. `http://www.gnu.org/software/gsl` (accessed: Dec. 15, 2008). Internet, 2008.
[101] H. H. González-Baños, D. Hsu, and J.-C. Latombe. Motion planning: Recent developments. In S. S. Ge and F. L. Lewis, editors, *Autonomous Mobile Robots: Sensing, Control, Decision-Making, and Applications*, chapter 10, pages 36–54. CRC Press, Boca Raton, 2006.
[102] D. M. Gorinevsky, A. M. Formalsky, and A. Y. Schneider. *Force Control of Robotics Systems.* CRC Press, Boca Raton, FL, USA, 1997.
[103] S. Haddadin, A. Albu-Schäffer, and G. Hirzinger. Safety evaluation of physical human-robot interaction via crash-testing. In *Proc. of Robotics: Science and Systems*, Atlanta, GA, USA, September 2007.
[104] S. Haddadin, A. Albu-Schäffer, and G. Hirzinger. The role of the robot mass and velocity in physical human-robot interaction - part I: Non-constrained blunt impacts. In *Proc. of the IEEE International Conference on Robotics and Automation*, pages 1331–1338, Passadena, CA, USA, May 2008.
[105] S. Haddadin, A. Albu-Schäffer, and G. Hirzinger. The role of the robot mass and velocity in physical human-robot interaction - part II: Constrained blunt impacts. In *Proc. of the IEEE International Conference on Robotics and Automation*, pages 1339–1345, Passadena, CA, USA, May 2008.
[106] S. Haddadin, A. Albu-Schäffer, A. De Luca, and G. Hirzinger. Collision detection and reaction: A contribution to safe physical human-robot interaction. In *Proc. of the IEEE/RSJ International Conference on Intelligent Robots and Systems*, pages 3356–3363, Nice, France, September 2008.
[107] Hasbro Inc., 1027 Newport Avenue, Mailstop A906, Pawtucket, RI 02861, USA. Jenga homepage. `http://www.jenga.com` (accessed: Dec. 15, 2008). Internet, 2008.
[108] R. Haschke, E. Weitnauer, and H. Ritter. On-line planning of time-optimal, jerk-limited trajectories. In *Proc. of the IEEE/RSJ International Conference on Intelligent Robots and Systems*, pages 3248–3253, Nice, France, September 2008.
[109] A. Heim and O. von Stryk. Trajectory optimization of industrial robots with application to computer-aided robotics and robot controllers. *Optimization*, 47:407–420, 2000.
[110] E. A. Henis and T. Flash. Mechanisms underlying the generation of averaged modified trajectories. *Biological Cybernetics*, 72(5):407–419, April 1995.

[111] Don Herbison-Evans. Finding real roots of quartics. Technical report, Sydney University of Technology, Department of Software Engineering, Sydney, Australia, 2005.

[112] G. Hirzinger, N. Sporer, A. Albu-Schäffer, M. Hähnle, R. Krenn, A. Pascucci, and M. Schedl. DLR's torque-controlled light weight robot III — Are we reaching the technological limits now? In *Proc. of the IEEE International Conference on Robotics and Automation*, volume 2, pages 1710–1716, Washington, D.C., USA, May 2002.

[113] N. Hogan. Impedance control: An approach to manipulation. Part I: Theory. Part II: Implementation. Part III: Applications. *ASME Journal of Dynamic Systems, Measurment, and Control*, 107:1–24, March 1985.

[114] J. M. Hollerbach. Dynamic scaling of manipulator trajectories. *ASME Journal on Dynamic Systems, Measurement, and Control*, 106(1):102–106, 1984.

[115] Institut für Robotik und Prozessinformatik at the Technische Universität Carolo-Wilhelmina zu Braunschweig, Mühlenpfordtstr. 23, D-38106 Braunschweig, Germany. Hompepage. `http://www.rob.tu-bs.de/en` (accessed: Dec. 15, 2008). Internet, 2008.

[116] International Federation of Robotics (IFR). Hompepage. `http://www.ifr.org` (accessed: Dec. 15, 2008). Internet, 2008.

[117] International Organisation for Standardisation. ISO 8373: Manipulating industrial robots — Vocabulary. International Standard, 1994.

[118] L. Jaillet and T. Siméon. A PRM-based motion planner for dynamically changing environments. In *Proc. of the IEEE/RSJ International Conference on Intelligent Robots and Systems*, volume 2, pages 1606–1611, Sendai, Japan, September 2004.

[119] JR3, Inc., 22 Harter Ave, Woodland, CA 95776, USA. Homepage. `http://www.jr3.com` (accessed: Dec. 15, 2008). Internet, 2008.

[120] M. E. Kahn and B. Roth. The near-minimum-time control of open-loop articulated kinematic chains. *ASME Journal of Dynamic Systems, Measurement, and Control*, 93:164–172, September 1971.

[121] C.-G. Kang. Online trajectory planning for a PUMA robot. *International Journal of Precision Engineering and Manufacturing*, 8(4):51–56, October 2007.

[122] L. E. Kavraki and S. M. LaValle. Motion planning. In B. Siciliano and O. Khatib, editors, *Springer Handbook of Robotics*, chapter 5, pages 109–131. Springer, Berlin, Heidelberg, Germany, first edition, 2008.

[123] Kawasaki Heavy Industries, Ltd., World Trade Center Bldg., 4-1, Hamamatsu-cho 2-chome, Minato-ku, Tokyo 105-6116, Japan. Homepage. `http://www.khi.co.jp/index_e.html` (accessed: Dec. 15, 2008). Internet, 2008.

[124] Kawasaki Robotics GmbH, Sperberweg 29, D-41468 Neuss, Germany. *Kawasaki Robot Controller der Serie D, Bedienungshandbuch (in German)*, 2002. 90209–1017DGB.

[125] Kawasaki Robotics GmbH, Sperberweg 29, D-41468 Neuss, Germany. *Kawasaki Robot Controller Serie D, Referenzhandbuch AS-Sprache (in German)*, 2002. 90209–1083DGE.

[126] W. Khalil and E. Dombre. *Modeling, Identification and Control of Robots*, chapter 13, Trajectory Generation, pages 313–345. Hermes Penton, Ltd., London, UK, first edition, 2002.

[127] W. Khalil and E. Dombre. *Modeling, Identification and Control of Robots*, chapter 14, Motion Control, pages 347–376. Hermes Penton, Ltd., London, UK, first edition, 2002.

[128] W. Khalil and E. Dombre. *Modeling, Identification and Control of Robots*. Hermes Penton, Ltd., London, UK, 2002.

[129] O. Khatib. A unified approach for motion and force control of robot manipulators: The operational space formulation. *IEEE Journal of Robotics and Automation*, RA-3(1):43–53, February 1987.

[130] J.-Y. Kim, D.-H. Kim, and S.-R. Kim. On-line minimum-time trajectory planning for industrial manipulators. In *Proc. of the International Conference on Control, Automation, and Systems*, pages 36–40, Seoul, South Korea, October 2007.

[131] R. King. An improved pegasus method for root finding. *BIT Numerical Mathematics*, 13(4):423–427, December 1973.

[132] B. Koninckx and H. van Brussel. Real-time NURBS interpolator for distributed motion control. *CIRP Annals*.

[133] Y. Koren. Control of machine tools. *Journal of Manufacturing Science and Engineering*, 119(4B):749–755, November 1997.

[134] A. I. Kostrikin, Y. I. Manin, and M. E. Alferieff. *Linear Algebra and Geometry*, chapter 3, Projective Groups and Projections. Taylor and Francis, Ltd., first edition, 1997.

[135] K. Kozłowski. *Modelling and Identification in Robotics*. Springer, London, UK, 1998.

[136] T. Kröger, B. Finkemeyer, M. Heuck, and F. M. Wahl. Adaptive implicit hybrid force/pose control of industrial manipulators: Compliant motion experiments. In *Proc. of the IEEE/RSJ International Conference on Intelligent Robots and Systems*, pages 816–821, Sendai, Japan, September 2004.

[137] T. Kröger, B. Finkemeyer, U. Thomas, and F. M. Wahl. Compliant motion programming: The task frame formalism revisited. In *Mechatronics and Robotics*, pages 1029–1034, Aachen, Germany, September 2004.

[138] T. Kröger, B. Finkemeyer, and F. M. Wahl. Manipulation primitives as interface between task programming and execution. In *Workshop on Issues and Approaches to Task Level Control at the IEEE/RSJ International Conference on Intellegent Robots and Systems*, Sendai, Japan, September 2004.

[139] T. Kröger, B. Finkemeyer, and F. M. Wahl. A task frame formalism for practical implementations. In *Proc. of the IEEE International Conference on Robotics and Automation*, pages 5218–5223, New Orleans, LA, USA, April 2004.

[140] T. Kröger, B. Finkemeyer, S. Winkelbach, S. Molkenstruck, L.-O. Eble, and F. M. Wahl. Demonstration of multi-sensor integration in industrial manipulation (poster). In *Proc. of the IEEE International Conference on Robotics and Automation*, pages 4282–4284, Orlando, FL, USA, May 2006.

[141] T. Kröger, B. Finkemeyer, S. Winkelbach, S. Molkenstruck, L.-O. Eble, and F. M. Wahl. Demonstration of multi-sensor integration in industrial manipulation (video). In *Proc. of the IEEE International Conference on Robotics and Automation*, Orlando, FL, USA, May 2006.

[142] T. Kröger, B. Finkemeyer, S. Winkelbach, S. Molkenstruck, L.-O. Eble, and F. M. Wahl. A manipulator plays Jenga. *IEEE Robotics and Automation Magazine*, 15(3):79–84, September 2008.

[143] T. Kröger, D. Kubus, and F. M. Wahl. 6D force and acceleration sensor fusion for compliant manipulation control. In *Proc. of the IEEE/RSJ International Conference on Intelligent Robots and Systems*, pages 2626–2631, Beijing, China, October 2006.

[144] T. Kröger, D. Kubus, and F. M. Wahl. Force and acceleration sensor fusion for compliant manipulation control in six degrees of freedom. *Advanced Robotics*, 21(14):1603–1616, November 2007.

[145] T. Kröger, D. Kubus, and F. M. Wahl. 12d force and acceleration sensing: A helpful experience report on sensor characteristics. In *Proc. of the IEEE International Conference on Robotics and Automation*, pages 3455–3462, Pasadena, CA, USA, May 2008.

[146] T. Kröger, A. Tomiczek, and F. M. Wahl. Towards on-line trajectory computation. In *Proc. of the IEEE/RSJ International Conference on Intelligent Robots and Systems*, pages 736–741, Beijing, China, October 2006.

[147] D. Kubus, T. Kröger, and F. M. Wahl. On-line rigid object recognition and pose estimation based on inertial parameters. In *Proc. of the IEEE/RSJ International Conference on Intelligent Robots and Systems*, pages 1402–1408, San Diego, CA, USA, October 2007.

[148] D. Kubus, T. Kröger, and F. M. Wahl. Improving force control performance by computational elimination of non-contact forces/torques. In *Proc. of the IEEE International Conference on Robotics and Automation*, pages 2617–2622, Pasadena, CA, USA, May 2008.

[149] A. Kugi, C. Ott A. Albu-Schäffer, and G. Hirzinger. On the passivity-based impedance control of flexible joint robots. *IEEE Trans. on Robotics*, 24(2):416–429, April 2008.

[150] KUKA Roboter GmbH, Zugspitzstraße 140, D-86165 Augsburg, Germany. *KUKA.Occubot VI V2.0 für KUKA System Software (KSS) V5.2 (in German)*, 2006. V0.1 24.01.2006 KST-AD-Occubot20 de.

[151] KUKA Roboter GmbH, Zugspitzstraße 140, D-86165 Augsburg, Germany. *KUKA System Software 5.2, 5.3, 5.4, Bedien- und Programmieranleitung für Systemintegratoren (in German)*, 2007. V1.1 05.07.2007 KSS-AD-SI-5x de.

[152] KUKA Roboter GmbH, Zugspitzstraße 140, D-86165 Augsburg, Germany. *KUKA KRC2 Brochure: Control and software*, 2008. WM-Nr. 841706/E/8/05.04.

[153] KUKA Roboter GmbH, Zugspitzstraße 140, D-86165 Augsburg, Germany. Homepage. `http://www.kuka.com/en/company/group` (accessed: Dec. 15, 2008). Internet, 2008.

[154] K. J. Kyriakopoulos and G. N. Sridis. Minimum jerk path generation. In *Proc. of the IEEE International Conference on Robotics and Automation*, volume 1, pages 364–369, Philadelphia, PA, USA, April 1988.

[155] P. Lambrechts, M. Boerlage, and M. Steinbuch. Trajectory planning and feedforward design for high performance motion systems. In *Proc. of the American Control Conference*, volume 5, pages 4637–4642, Boston, MA, USA, June 2004.

[156] S. M. LaValle. *Planning Algorithms*. Cambridge University Press, Cambridge, UK, 2006.

[157] W. T. Lei, M. P. Sung, L. Y. Lin, and J. J. Huang. Fast real-time NURBS for CNC machine tools. *International Journal of Machine Tools and Manufacture*, 47(10):1530–1541, 2007.

[158] Lenze AG, Hans-Lenze-Straße 1, D-31855 Aerzen, Germany. *Software manual, L-force Servo Drives 9400 StateLine, E94AxSExxxx, parameter setting*, 2008. DMS 3.0 EN - 07/2008.

[159] Lenze AG, Hans-Lenze-Straße 1, D-31855 Aerzen, Germany. Homepage. `http://www.lenze.com` (accessed: Dec. 15, 2008). Internet, 2008.

[160] W. Leonhard. *Control of Electrical Drives*. Springer, third edition, 2001.

[161] T. Y. Li and J.-C. Latombe. On-line manipulation planning for two robot arms in a dynamic environment. In *Proc. of the IEEE International Conference on Robotics and Automation*, volume 1, pages 1048–1055, Nagoya, Japan, May 1995.

[162] T. Y. Li and J.-C. Latombe. On-line manipulation planning for two robot arms in a dynamic environment. *The International Journal of Robotics Research*, 16(2):144–167, April 1997.

[163] D. Liberzon. *Switching in Systems and Control*. Systems and Control: Foundations and Applications. Birkhäuser, Boston, MA, USA, 2003.

[164] D. Liberzon and A. S. Morse. Basic problems in stability and design of switched systems. *IEEE Control Systems Magazine*, 19(5):59–70, October 1999.

[165] C.-S. Lin, P.-R. Chang, and J. Y. S. Luh. Formulation and optimization of cubic polynomial joint trajectories for industrial robots. *IEEE Trans. on Automatic Control*, 28(12):1066–1074, December 1983.

[166] S. Lindemann and S. LaValle. Current issues in sampling-based motion planning. In P. Dario and R. Chatila, editors, *Proc. of the Eighth Int. Symp. on Robotics Research*, pages 36–54. Springer, Berlin, Germany, 2004.

[167] S. Liu. An on-line reference-trajectory generator for smooth motion of impulse-controlled industrial manipulators. In *Proc. of the seventh International Workshop on Advanced Motion Control*, pages 365–370, Maribor, Slovenia, July 2002.

[168] J. Lloyd and V. Hayward. Trajectory generation for sensor-driven and time-varying tasks. *The International Journal of Robotics Research*, 12(4):380–393, August 1993.

[169] T. Lozano-Pérez. Automatic planning of manipulator transfer movements. In M. Brady, J. M. Hollerbach, T. L. Johnson, T. Lozano-Pérez, and M. T. Mason, editors, *Robot Motion: Planning and Control*, chapter 6, pages 499–535. MIT Press, 1982.

[170] A. De Luca, A. Albu-Schäffer, S. Haddadin, and G. Hirzinger. Collision detection and safe reaction with the DLR-III lightweight manipulator arm. In *Proc. of the IEEE/RSJ International Conference on Intelligent Robots and Systems*, pages 1623–1630, Beijing, China, October 2006.

[171] Y. Ma, S. Soatto, J. Košecká, and S. S. Sastry. *An Invitation to 3-D Vision*. Springer, New York, NY, USA, 2003.

[172] J. Maaß, S. Molkenstruck, U. Thomas, J. Hesselbach, F. M. Wahl, and A. Raatz. Definition and execution of a generic assembly programming paradigm. *Assembly Automation*, 28(1):61–68, 2008.

[173] J. Maaß, J. Steiner, A. Raatz, J. Hesselbach, U. Goltz, and A. Amado. Self-management in a control architecture for parallel kinematic robots. In *Proc. of the 27th ASME Computers and Information in Engineering Conference*, New York, NY, USA, August 2008.

[174] S. Macfarlane and E. A. Croft. Jerk-bounded manipulator trajectory planning: Design for real-time applications. *IEEE Trans. on Robotics and Automation*, 19(1):42–52, February 2003.

[175] manutec VaWe Robotersystem GmbH, Benno-Strauß-Straße 5, D-90763 Fürth, Germany. Homepage. `http://www.manutec.de` (accessed: Dec. 15, 2008). Internet, 2008.

[176] M. T. Mason. Compliance and force control for computer controlled manipulators. *IEEE Trans. on Systems, Man, and Cybernetics*, 11:418–432, June 1981.

[177] J. B. Mbede, L. Zhang, S. Ma, Y. Toure, and V. Graefe. Robust neuro-fuzzy manipulator among navigation of mobile dynamic obstacles. In *Proc. of the IEEE International Conference on Robotics and Automation*, pages 5051–5057, New Orleans, LA, USA, April 2004.

[178] J. M. McCarthy and J. E. Bobrow. The number of saturated actuators and constraint forces during time-optimal movement of a general robotic system. *IEEE Trans. on Robotics and Automation*, 8(3):407–409, June 1992.

[179] N. H. McClamroch and I. Kolmanobsky. Performance benefits of hybrid control design for linear and nonlinear systems. *Proc. of the IEEE*, 88(7):1083–1096, July 2000.

[180] E. A. Merchán-Cruz and A. S. Morris. Fuzzy-GA-based trajectory planner for robot manipulators sharing a common workspace. *IEEE Trans. on Robotics and Automation*, 22(4):613–624, August 2006.

[181] G. Michaletzky and L. Gerencsér. BIBO stability of linear switching systems. *IEEE Trans. on Automatic Control*, 47(11):1895–1898, November 2002.

[182] G. Milighetti and H.-B. Kuntze. On a primitive skill-based supervisory robot control architecture. In *Proc. of the IEEE International Conference on Advanced Robotics*, pages 141–147, Seattle, WA, USA, July 2005.

[183] G. Milighetti and H.-B. Kuntze. On the discrete-continuous control of basic skills for humanoid robots. In *Proc. of the IEEE/RSJ International Conference on Intelligent Robots and Systems*, pages 3474–3479, Beijing, China, October 2006.

[184] G. Milighetti and H.-B. Kuntze. Fuzzy based decision making for the discrete-continuous control of humanoid robots. In *Proc. of the IEEE/RSJ International Conference on Intelligent Robots and Systems*, pages 3580–3585, San Diego, CA, USA, October 2007.

[185] Mitsubishi Heavy Industries Industrial Machinery Co., Ltd., 1-Takamichi, Iwatsuka-cho, Nakamura-ku, Nagoya, Japan. Homepage. `http://robot.sanki-mhi.com/English` (accessed: Dec. 15, 2008). Internet, 2008.

[186] Mitsubishi Heavy Industries, Ltd., Mitsubishijuko Yokohama Bldg., 3-1, Minatomirai 3-chome, Nishi-ku, Yokohama, 220-8401, Japan. Homepage. `http://www.mhi.co.jp/en` (accessed: Dec. 15, 2008). Internet, 2008.

[187] H. Mosemann. *Beiträge zur Planung, Dekomposition und Ausführung von automatisch generierten Roboteraufgaben (in German)*. Shaker Verlag, Aachen, Germany, 2000.

[188] H. Mosemann and F. M. Wahl. Automatic decomposition of planned assembly sequences into skill primitives. *IEEE Trans. on Robotics and Automation*, 17(5):709–718, October 2001.

[189] MOTOMAN, Inc., 805 Liberty Lane, West Carrollton, Ohio 45449, USA. Homepage. `http://www.motoman.com` (accessed: Dec. 15, 2008). Internet, 2008.

[190] MOTOMAN robotec GmbH, Kammerfeldstraße 1, D-85391 Allershausen, Germany. *System-Setup MOTOMAN NX100, Betriebsanleitung*, 2004. MRS6101GB.0.U.

[191] MOTOMAN robotec GmbH, Kammerfeldstraße 1, D-85391 Allershausen, Germany. *Betriebsanleitung Grundprogrammierung*, 2008. MRS60000.

[192] Neuronics AG, Technoparkstrasse 1, CH-8005 Zürich, Switzerland. *Data sheet Kantana — Automation made easy*, 2008.

[193] Neuronics AG, Technoparkstrasse 1, CH-8005 Zürich, Switzerland. *Katana-NativeInterface Reference Manual, Version 3.9.x*, 2008.

[194] Neuronics AG, Technoparkstrasse 1, CH-8005 Zürich, Switzerland. Homepage. `http://www.neuronics.ch` (accessed: Dec. 15, 2008). Internet, 2008.

[195] P. Ögren, M. Egerstedt, and X. Hu. Reactive mobile manipulation using dynamic trajectory tracking. In *Proc. of the IEEE International Conference on Robotics and Automation*, pages 3473–3478, San Francisco, CA, USA, April 2000.

[196] J. Olomski. *Bahnplanung und Bahnführung von Industrierobotern (in German)*. Vieweg Verlag, Braunschweig, Germany, 1989.

[197] OROCOS Homepage. Open robot control software. `http://www.orocos.org` (accessed: Dec. 15, 2008). Internet, 2002.

[198] R. Osypiuk, T. Kröger, B. Finkemeyer, and F. M. Wahl. A two-loop implicit force/position control structure, based on a simple linear model: Theory and experiment. In *Proc. of the IEEE International Conference on Robotics and Automation*, pages 2232–2237, Orlando, FL, USA, May 2006.

[199] W. S. Owen, E. A. Croft, and B. Benhabib. Minimally compliant trajectory resolution for robotic machining. In *Proc. of the IEEE International Conference on Advanced Robotics*, pages 702–707, Coimbra, Portugal, June 2003.

[200] W. S. Owen, E. A. Croft, and B. Benhabib. Real-time trajectory resolution for dual robot machining. In *Proc. of the IEEE/RSJ International Conference on Intelligent Robots and Systems*, pages 4332–4337, Sendai, Japan, September 2004.

[201] F. C. Park, J. E. Bobrow, and S. R. Ploen. A lie group formulation of robot dynamics. *The International Journal of Robotics Research*, 14(6):609–618, December 1995.

[202] F. C. Park and B. Ravani. Smooth invariant interpolation of rotations. *ACM Transactions on Graphics,*, 16(3):277–295, July 1997.

[203] R. P. C. Paul. Manipulator Cartesian path control. In M. Brady, J. M. Hollerbach, T. L. Johnson, T. Lozano-Pérez, and M. T. Mason, editors, *Robot Motion: Planning and Control*, chapter 4, pages 245–263. MIT Press, 1982.

[204] R. P. C. Paul. *Robot Manipulators*. MIT Press, fifth edition, 1983.

[205] R. P. C. Paul. *Robot Manipulators*, chapter 5, Motion Trajectories, pages 119–155. MIT Press, fifth edition, 1983.

[206] R. P. C. Paul. *Robot Manipulators*, chapter 7, Control, pages 197–215. MIT Press, fifth edition, 1983.

[207] F. Pertin and J.-M. Bonnet des Tuves. Real time robot controller abstraction layer. In *Proc. of the Int. Symposium on Robots*, Paris, France, March 2004.

[208] F. Pfeiffer and R. Johanni. A concept for manipulator trajectory planning. In *Proc. of the IEEE International Conference on Robotics and Automation*, volume 3, pages 1399–1405, San Francisco, CA, USA, April 1986.

[209] A. Piazzi and A. Visioli. Global minimum-time trajectory planning of mechanical manipulators using interval analysis. *International Journal of Control*, 71(4):631–652, 1998.

[210] A. Piazzi and A. Visioli. Global minimum-jerk trajectory planning of robot manipulators. *IEEE Trans. on Industrial Electronics*, 47(1):140–149, February 2000.

[211] L. Piegel. On NURBS: A survey. *IEEE Computer Graphics and Applications*, 11(1):55–71, January 1991.

[212] L. Piegel and W. Tiller. *The NURBS Book.* Springer, Heidelberg, Germany, second edition, 1997.

[213] F. Pierrot, P. Dauchez, and A. Fournier. HEXA: A fast six-DOF fully-parallel robot. In *Proc. of the IEEE International Conference on Advanced Robotics*, volume 2, pages 1158–1163, Pisa, Italy, June 1991.

[214] I. Pietsch. *Adaptive Steuerung und Regelung ebener Parallelroboter (in German).* Vulkan Verlag, Essen, Germany, 2003.

[215] W. H. Press, S. A. Teukolski, W. T. Vetterling, and B. P. Flannery. *Numerical Receipes in C++.* Cambridge University Press, 2002.

[216] QNX Software Systems, 175 Terence Matthews Crescent, Ottawa, Ontario, Canada, K2M 1W8. Homepage. `http://www.qnx.com` (accessed: Dec. 15, 2008). Internet, 2008.

[217] M. H. Raibert and J. J. Craig. Hybrid position/force control of manipulators. *ASME Journal of Dynamic Systems, Measurement and Control*, 102:126–133, June 1981.

[218] V. T. Rajan. Minimum time trajectory planning. In *Proc. of the IEEE International Conference on Robotics and Automation*, volume 3, pages 759–764, St. Louis, MO, USA, March 1985.

[219] Real-Time Innovations, 385 Moffett Park Drive, Sunnyvale, CA 94089, USA. Homepage. `http://www.rti.com` (accessed: Dec. 15, 2008). Internet, 2008.

[220] T. Reisinger. *Kontaktregelung von Parallelrobotern auf der Basis von Aktionsprimitiven (Interaction Control of Parallel Robots Based on Skill Primitives).* PhD thesis, Institut für Regelungstechnik, Technische Universität Carolo-Wilhelmina zu Braunschweig, `http://www.digibib.tu-bs.de/?docid=00022368` (accessed: Dec. 15, 2008), 2008.

[221] Rockwell Automation, Inc., Allen-Bradley and Rockwell Software Brands, 1201 South Second Street, Milwaukee, WI 53204-2496, USA. Homepage. `http://www.rockwellautomation.com` (accessed: Dec. 15, 2008). Internet, 2008.

[222] V. Rogozin, Y. Edan, and T. Flash. A real-time trajectory modification algorithm. *Robotica*, 19(4):395–405, July 2001.

[223] G. Sahar and J. M. Hollerbach. Planning of minimum-time trajectories for robot arms. In *Proc. of the IEEE International Conference on Robotics and Automation*, volume 2, pages 751–758, St. Louis, MO, USA, March 1985.

[224] M. Schlemmer and G. Gruebel. Real-time collision-free trajectory optimization of robot manipulators via semi-infinite parameter optimization. *The International Journal of Robotics Research*, 17(9):1013–1021, September 1998.

[225] D. C. Schmidt, D. L. Levine, and S. Mungee. The design of the TAO real-time object request broker. *Computer Communications*, 21(4):294–324, April 1998.

[226] J. De Schutter, T. De Laet, J. Rutgeerts, W. Decré, R. Smits, E. Aertbeliën, K. Claes, and H. Bruyninckx. Constraint-based task specification and estimation for sensor-based robot systems in the presence of geometric uncertainty. *The International Journal of Robotics Research*, 26(5):433–454, May 2007.

[227] J. De Schutter and J. van Brussel. Compliant robot motion I. A formalism for specifying compliant motion tasks. *The International Journal of Robotics Research*, 7(5):3–17, August 1988.

[228] J. De Schutter and J. van Brussel. Compliant robot motion II. A control approach based on external control loops. *The International Journal of Robotics Research*, 7(4):18–33, August 1988.

[229] L. Sciavicco and B. Siciliano. *Modelling and Control of Robot Manipulators*. Advanced Textbooks in Control and Signal Processing. Springer, second edition, 2000.

[230] A. I. Selverston, M. I. Rabinovich, R. Huerta, T. Novotny, R. Levi, Y. Axshavsky, A. Volkovskii, J. Ayers, and R. Pinto. Biomimetic central pattern generators for robotics and prosthetics. In *Proc. of the IEEE International Conference on Robotics and Biomimetics*, pages 885–888, Chenyang, China, August 2004.

[231] SERCOS International e.V., Küblerstrasse 1, D-73079 Süssen, Germany. Homepage. `http://www.sercos.de` (accessed: Dec. 15, 2008). Internet, 2008.

[232] L. E. Sergio, C. Hamel-Pâquet, and J. F. Kalaska. Motor cortex neural correlates of output kinematics and kinetics during isometric-force and arm-reaching tasks. *Journal of Neurophysiology*, 94:2353–2378, May 2005.

[233] SEW-EURODRIVE GmbH & Co KG, Ernst-Blickle-Straße 42, D-76646 Bruchsal, Germany. Homepage. `http://www.sew-eurodrive.com` (accessed: Dec. 15, 2008). Internet, 2008.

[234] Z. Shiller. Time-energy optimal control of articulated systems with geometric path constraints. In *Proc. of the IEEE International Conference on Robotics and Automation*, volume 4, pages 2680–2685, San Diego, CA, USA, May 1994.

[235] Z. Shiller and S. Dubowski. Time optimal paths and acceleration lines of robotic manipulators. In *Proc. of the IEEE Conference on Decision and Control*, volume 26, pages 199–204, Los Angeles, CA, USA, December 1987.

[236] Z. Shiller and H.-H. Lu. Robust computation of path constrained time optimal motions. In *Proc. of the IEEE International Conference on Robotics and Automation*, volume 1, pages 144–149, Cincinnati, OH, USA, May 1990.

[237] Z. Shiller and H.-H. Lu. Computation of path constrained time optimal motions with dynamic singularities. *ASME Journal of Dynamic Systems, Measurement, and Control*, 114(1):34–40, March 1992.

[238] K. G. Shin and N. D. McKay. Minimum-time control of robotic manipulators with geometric path constraints. *IEEE Trans. on Automatic Control*, 30(5):531–541, June 1985.

[239] K. G. Shin and N. D. McKay. A dynamic programming approach to trajectory planning of robotic manipulators. *IEEE Trans. on Automatic Control*, 31(6):491–500, June 1986.

[240] B. Siciliano and O. Khatib. *Springer Handbook of Robotics.* Springer, first edition, 2008.

[241] B. Siciliano and L. Villani. *Robot Force Control.* Kluwer Academic Publishers, 1999.

[242] Siemens AG, Wittelsbacherplatz 2, D-80333 Munich, Germany. Homepage. `http://www.automation.siemens.com` (accessed: Dec. 15, 2008). Internet, 2008.

[243] D. Simon and C. Isik. A trigonometric trajectory generator for robotic arms. *International Journal of Control*, 57(3):505–517, March 1993.

[244] J.-J. E. Slotine and H. S. Yang. Improving the efficiency of time-optimal path-following algorithms. *IEEE Trans. on Robotics and Automation*, 5(1):118–124, February 1989.

[245] A. J. Sommese and C. W. Wampler. *The Numerical Solution of Systems of Polynomials.* World Scientific, Singapore, first edition, 2005.

[246] M. W. Spong, S. A. Hutchinson, and M. Vidyasagar. *Robot Modeling and Control.* John Wiley and Sons, 2006.

[247] Stäubli Faverges SCA, Place Robert Stäubli BP 70, 74210 Faverges (Annecy), France. *Documentation CS8 Low Level Interface (LLI) s5.3.2*, 2006. Version D24276405A.

[248] Stäubli Faverges SCA, Place Robert Stäubli BP 70, 74210 Faverges (Annecy), France. *Instruction Manual CS8C Controller*, 2008. D28062904A-01/2006.

[249] Stäubli Faverges SCA, Place Robert Stäubli BP 70, 74210 Faverges (Annecy), France. *VAL3 Reference Manual Version 5.3*, 2008. D28062804A-02/2006.

[250] Stäubli Faverges SCA, Place Robert Stäubli BP 70, 74210 Faverges (Annecy), France. Homepage. `http://www.staubli.com/en/robotics` (accessed: Dec. 15, 2008). Internet, 2008.

[251] J. Stoer and R. Bulrisch. *Introduction to Numerical Analysis.* Springer, New York, NY, USA, 2002.

[252] O. von Stryk and M. Schlemmer. Optimal control of the industrial robot manutec r3. In R. Burlish and D. Kraft, editors, *Computational Optimal Control, International Series of Numerical Mathematics*, volume 115, pages 367–382, Basel, 1994. Birkhäuser.

[253] S.-H. Suh, S.-K. Kang, D.-H. Chung, and I. Stroud. *Theory and Design of CNC Systems.* Springer Series in Advanced Manufacturing. Springer, London, UK, 2008.

[254] K. Takayama and H. Kano. A new approach to synthesizing free motions of robotic manipulators based on a concept of unit motions. In *Proc. of the IEEE/RSJ International Conference on Intellegent Robots and Systems, International Workshop on Intelligence for Mechanical Systems*, volume 3, pages 1406–1411, Osaka, Japan, November 1991.

[255] H. H. Tan and R. B. Potts. Minimum time trajectory planner for the discrete dynamic robotmodel with dynamic constraints. *IEEE Trans. on Robotics and Automation*, 4(2):174–184, April 1988.

[256] R. H. Taylor. Planning and execution of straight-line manipulator trajectories. In M. Brady, J. M. Hollerbach, T. L. Johnson, T. Lozano-Pérez, and M. T. Mason, editors, *Robot Motion: Planning and Control*, chapter 4, pages 265–286. MIT Press, 1982.

[257] The Numerical Algorithms Group, Ltd., Wilkinson House Jordan Hill Road, Oxford, OX2 8DR, UK. Hompepage. `http://www.nag.co.uk` (accessed: Dec. 15, 2008). Internet, 2008.

[258] The University of Chicago. *The Chicago Manual of Style*. The University of Chicago Press, Chicago, IL, USA, 15th edition, 2003.

[259] U. Thomas. *Automatisierte Programmierung von Robotern für Montageaufgaben (in German)*. Shaker Verlag, Aachen, Germany, 2008.

[260] U. Thomas, B. Finkemeyer, T. Kröger, and F. M. Wahl. Error-tolerant execution of complex robot tasks based on skill primitives. In *Proc. of the IEEE International Conference on Robotics and Automation*, volume 3, pages 3069–3075, Taipei, Taiwan, September 2003.

[261] U. Thomas, F. M. Wahl, J. Maaß, and J. Hesselbach. Towards a new concept of robot programming in high speed assembly applications. In *Proc. of the IEEE/RSJ International Conference on Intelligent Robots and Systems*, pages 3827–3833, Edmonton, Canada, August 2005.

[262] G. J. Tortora and B. Derrickson. *Principles of Anatomy and Physiology*, chapter 13, pages 439–472. John Wiley and Sons, eleventh edition, 2006.

[263] G. J. Tortora and B. Derrickson. *Principles of Anatomy and Physiology*. John Wiley and Sons, eleventh edition, 2006.

[264] J. F. Traub. *Iterative Methods for the Solution of Equations*, chapter 1.2 and appendix C. Prentice-Hall, Englewood Cliffs, NJ, USA, first edition, 1964.

[265] I. R. van Aken and H. van Brussel. On-line robot trajectory control in joint coordinates by means of imposed acceleration profiles. *Robotica*, 6(3):185–195, 1988.

[266] J. Vannoy and J. Xiao. Real-time adaptive motion planning (RAMP) of mobile manipulators in dynamic environments with unforeseen changes. *IEEE Trans. on Robotics*, 24(5):1199–1212, October 2008.

[267] L. Villani, , and J. De Schutter. Force control. In B. Siciliano and O. Khatib, editors, *Springer Handbook of Robotics*, chapter 7, pages 161–185. Springer, Berlin, Heidelberg, Germany, first edition, 2008.

[268] M. Vukobratović and D. Šurdilović. Control of robotic systems in contact tasks: An overview. In *Proc. of the IEEE International Conference on Robotics and Automation*, pages 13–32, Atlanta, GA, USA, May 1993.

[269] M. Žefran, V. Kumar, and C. B. Croke. On the generation of smooth three-dimensional rigid body motions. *IEEE Trans. on Robotics and Automation*, 14(4):576–589, August 1998.

[270] L. Žlajpah. On time optimal path control of manipulators with bounded joint velocities and torques. In *Proc. of the IEEE International Conference on Robotics and Automation*, volume 2, pages 1572–1577, Minneapolis, MN, USA, April 1996.

[271] F. M. Wahl. *Digital Image Signal Processing*. Artech House, Boston, MA, USA, first edition, 1987.

[272] W. Wang, S. S. Chan, D. A. Heldman, and D. W. Moran. Motor cortical representation of position and velocity during reaching. *Journal of Neurophysiology*, 97:4258–4270, March 2007.

[273] L. E. Weiss, A. C. Sanderson, and C.P. Neuman. Dynamic sensor-based control of robots with visual feedback. *IEEE Journal of Robotics and Automation*, 3(5):404–417, 1987.

[274] D. E. Whitney. Resolved motion rate control of manipulators and human prostheses. *IEEE Trans. on Man Machine Systems*, 10(2):47–53, June 1969.
[275] D. E. Whitney. Force feedback control of manipulator fine motion. *ASME Journal of Dynamic Systems, Measurment and Control*, 98:91–97, 1977.
[276] D. E. Whitney and J. L. Nevins. What is the remote center compliance (RCC) and what can it do? In *Proc. of the ninth International Symposium on Industrial Robots*, pages 135–152, Washington, D.C., USA, March 1979.
[277] M. A. Wicks, P. Peleties, and R. A. DeCarlo. Construction of piecewise Lyapunov functions for stabilizing switched systems. In *Proc. of the IEEE Conference on Decision and Control*, volume 4, pages 3492–3497, Lake Buena Vista, FL, USA, December 1994.
[278] C. C. De Wit, B. Siciliano, and G. Bastin. *Theory of Robot Control.* Springer, 1996.
[279] X. Xu and G. Zhai. Practical stability and stabilization of hybrid and switched systems. *IEEE Trans. on Automatic Control*, 50(11):1897–1903, November 2005.
[280] Y. Yang and O. Brock. Elastic roadmaps: Globally task-consistent motion for autonomous mobile manipulation in dynamic environments. In *Proc. of Robotics: Science and Systems*, Philadelphia, PA, USA, August 2006.
[281] YASKAWA Electric Corporation, 2-1 Kurosaki-Shiroishi, Yahatanishi-Ku, Kitakyushu, Fukuoka 806-0004, Japan. Homepage. `http://www.yaskawa.com` (accessed: Dec. 15, 2008). Internet, 2008.
[282] R. Zaier and S. Kanda. Piecewise-linear pattern generator and reflex system for humanoid robots. In *Proc. of the IEEE International Conference on Robotics and Automation*, pages 2188–2195, Roma, Italy, April 2007.
[283] R. Zaier and F. Nagashima. Motion pattern generator and reflex system for humanoid robots. In *Proc. of the IEEE/RSJ International Conference on Intelligent Robots and Systems*, pages 840–845, Beijing, China, October 2006.

Springer Tracts in Advanced Robotics

Edited by B. Siciliano, O. Khatib and F. Groen

Further volumes of this series can be found on our homepage: springer.com

Vol. 57: Chirikjian, G.S.; Choset, H.; Morales, M., Murphey, T. (Eds.)
Algorithmic Foundations of Robotics VIII – Selected Contributions of the Eighth International Workshop on the Algorithmic Foundations of Robotics
680 p. 2010 [978-3-642-00311-0]

Vol. 56: Buehler, M.; Iagnemma, K.; Singh S. (Eds.)
The DARPA Urban Challenge – Autonomous Vehicles in City Traffic
625 p. 2009 [978-3-642-03990-4]

Vol. 55: Stachniss, C.
Robotic Mapping and Exploration
196 p. 2009 [978-3-642-01096-5]

Vol. 54: Khatib, O.; Kumar, V.; Pappas, G.J. (Eds.)
Experimental Robotics: The Eleventh International Symposium
579 p. 2009 [978-3-642-00195-6]

Vol. 53: Duindam, V.; Stramigioli, S.
Modeling and Control for Efficient Bipedal Walking Robots
211 p. 2009 [978-3-540-89917-4]

Vol. 52: Nüchter, A.
3D Robotic Mapping
201 p. 2009 [978-3-540-89883-2]

Vol. 51: Song, D.
Sharing a Vision
186 p. 2009 [978-3-540-88064-6]

Vol. 50: Alterovitz, R.; Goldberg, K.
Motion Planning in Medicine: Optimization and Simulation Algorithms for Image-Guided Procedures
153 p. 2008 [978-3-540-69257-7]

Vol. 49: Ott, C.
Cartesian Impedance Control of Redundant and Flexible-Joint Robots
190 p. 2008 [978-3-540-69253-9]

Vol. 48: Wolter, D.
Spatial Representation and Reasoning for Robot Mapping
185 p. 2008 [978-3-540-69011-5]

Vol. 47: Akella, S.; Amato, N.; Huang, W.; Mishra, B.; (Eds.)
Algorithmic Foundation of Robotics VII
524 p. 2008 [978-3-540-68404-6]

Vol. 46: Bessière, P.; Laugier, C.; Siegwart R. (Eds.)
Probabilistic Reasoning and Decision Making in Sensory-Motor Systems
375 p. 2008 [978-3-540-79006-8]

Vol. 45: Bicchi, A.; Buss, M.; Ernst, M.O.; Peer A. (Eds.)
The Sense of Touch and Its Rendering
281 p. 2008 [978-3-540-79034-1]

Vol. 44: Bruyninckx, H.; Přeučil, L.; Kulich, M. (Eds.)
European Robotics Symposium 2008
356 p. 2008 [978-3-540-78315-2]

Vol. 43: Lamon, P.
3D-Position Tracking and Control for All-Terrain Robots
105 p. 2008 [978-3-540-78286-5]

Vol. 42: Laugier, C.; Siegwart, R. (Eds.)
Field and Service Robotics
597 p. 2008 [978-3-540-75403-9]

Vol. 41: Milford, M.J.
Robot Navigation from Nature
194 p. 2008 [978-3-540-77519-5]

Vol. 40: Birglen, L.; Laliberté, T.; Gosselin, C.
Underactuated Robotic Hands
241 p. 2008 [978-3-540-77458-7]

Vol. 39: Khatib, O.; Kumar, V.; Rus, D. (Eds.)
Experimental Robotics
563 p. 2008 [978-3-540-77456-3]

Vol. 38: Jefferies, M.E.; Yeap, W.-K. (Eds.)
Robotics and Cognitive Approaches to Spatial Mapping
328 p. 2008 [978-3-540-75386-5]

Vol. 37: Ollero, A.; Maza, I. (Eds.)
Multiple Heterogeneous Unmanned Aerial Vehicles
233 p. 2007 [978-3-540-73957-9]

Vol. 36: Buehler, M.; Iagnemma, K.; Singh, S. (Eds.)
The 2005 DARPA Grand Challenge – The Great Robot Race
520 p. 2007 [978-3-540-73428-4]

Vol. 35: Laugier, C.; Chatila, R. (Eds.)
Autonomous Navigation in Dynamic Environments
169 p. 2007 [978-3-540-73421-5]

Vol. 34: Wisse, M.; van der Linde, R.Q.
Delft Pneumatic Bipeds
136 p. 2007 [978-3-540-72807-8]

Vol. 33: Kong, X.; Gosselin, C.
Type Synthesis of Parallel Mechanisms
272 p. 2007 [978-3-540-71989-2]

Vol. 30: Brugali, D. (Ed.)
Software Engineering for Experimental Robotics
490 p. 2007 [978-3-540-68949-2]

Vol. 29: Secchi, C.; Stramigioli, S.; Fantuzzi, C.
Control of Interactive Robotic Interfaces – A Port-Hamiltonian Approach
225 p. 2007 [978-3-540-49712-7]

Vol. 28: Thrun, S.; Brooks, R.; Durrant-Whyte, H. (Eds.)
Robotics Research – Results of the 12th International Symposium ISRR
602 p. 2007 [978-3-540-48110-2]

Vol. 27: Montemerlo, M.; Thrun, S.
FastSLAM – A Scalable Method for the Simultaneous Localization and Mapping Problem in Robotics
120 p. 2007 [978-3-540-46399-3]

Vol. 26: Taylor, G.; Kleeman, L.
Visual Perception and Robotic Manipulation – 3D Object Recognition, Tracking and Hand-Eye Coordination
218 p. 2007 [978-3-540-33454-5]

Vol. 25: Corke, P.; Sukkarieh, S. (Eds.)
Field and Service Robotics – Results of the 5th International Conference
580 p. 2006 [978-3-540-33452-1]

Vol. 24: Yuta, S.; Asama, H.; Thrun, S.; Prassler, E.; Tsubouchi, T. (Eds.)
Field and Service Robotics – Recent Advances in Research and Applications
550 p. 2006 [978-3-540-32801-8]

Vol. 23: Andrade-Cetto, J,; Sanfeliu, A.
Environment Learning for Indoor Mobile Robots – A Stochastic State Estimation Approach to Simultaneous Localization and Map Building
130 p. 2006 [978-3-540-32795-0]

Vol. 22: Christensen, H.I. (Ed.)
European Robotics Symposium 2006
209 p. 2006 [978-3-540-32688-5]

Vol. 21: Ang Jr., H.; Khatib, O. (Eds.)
Experimental Robotics IX – The 9th International Symposium on Experimental Robotics
618 p. 2006 [978-3-540-28816-9]

Vol. 20: Xu, Y.; Ou, Y.
Control of Single Wheel Robots
188 p. 2005 [978-3-540-28184-9]

Vol. 19: Lefebvre, T.; Bruyninckx, H.; De Schutter, J. Nonlinear Kalman Filtering for Force-Controlled Robot Tasks
280 p. 2005 [978-3-540-28023-1]

Vol. 18: Barbagli, F.; Prattichizzo, D.; Salisbury, K. (Eds.)
Multi-point Interaction with Real and Virtual Objects
281 p. 2005 [978-3-540-26036-3]

Vol. 17: Erdmann, M.; Hsu, D.; Overmars, M.; van der Stappen, F.A (Eds.)
Algorithmic Foundations of Robotics VI
472 p. 2005 [978-3-540-25728-8]

Vol. 16: Cuesta, F.; Ollero, A.
Intelligent Mobile Robot Navigation
224 p. 2005 [978-3-540-23956-7]

Vol. 15: Dario, P.; Chatila R. (Eds.)
Robotics Research – The Eleventh International Symposium
595 p. 2005 [978-3-540-23214-8]

Vol. 14: Prassler, E.; Lawitzky, G.; Stopp, A.; Grunwald, G.; Hägele, M.; Dillmann, R.; Iossifidis. I. (Eds.)
Advances in Human-Robot Interaction
414 p. 2005 [978-3-540-23211-7]

Vol. 13: Chung, W.
Nonholonomic Manipulators
115 p. 2004 [978-3-540-22108-1]

MIX
Papier aus verantwortungsvollen Quellen
Paper from responsible sources
FSC® C105338

If you have any concerns about our products,
you can contact us on
ProductSafety@springernature.com

In case Publisher is established outside the EU,
the EU authorized representative is:
Springer Nature Customer Service Center GmbH
Europaplatz 3, 69115 Heidelberg, Germany

Printed by Libri Plureos GmbH
in Hamburg, Germany